Back Analysis in Rock Engineering

ISRM Book Series
Series editor: Xia-Ting Feng
Institute of Rock and Soil Mechanics, Chinese Academy of Sciences, Wuhan, China

ISSN : 2326-6872
eISSN: 2326-778X

Volume 4

International Society for Rock Mechanics

ISRM

Back Analysis in Rock Engineering

Shunsuke Sakurai

Kobe University, Kobe, Japan

CRC Press

Taylor & Francis Group

Boca Raton London New York

CRC Press is an imprint of the
Taylor & Francis Group, an **informa** business

A BALKEMA BOOK

Published by:
CRC Press/Balkema
P.O. Box 447, 2300 AK Leiden, The Netherlands
e-mail: Pub.NL@taylorandfrancis.com
www.crcpress.com – www.taylorandfrancis.com

First issued in paperback 2021

Typeset by MPS Limited, Chennai, India

ISBN 13: 978-1-03-209650-6 (pbk)
ISBN 13: 978-1-138-02862-3 (hbk)

Publisher's Note
The publisher has gone to great lengths to ensure the quality of this reprint but points out that some imperfections in the original copies may be apparent.

Visit the Taylor & Francis Web site at
http://www.taylorandfrancis.com

and the CRC Press Web site at
http://www.crcpress.com

Library of Congress Cataloging-in-Publication Data

Names: Sakurai, Shunsuke, 1935– author.
Title: Back analysis in rock engineering / Shunsuke Sakurai, Kobe University, Kobe, Japan.
Description: Leiden, The Netherlands : CRC Press/Balkema, [2017] | Series: ISRM book series ; volume 4 | Includes bibliographical references and index.
Identifiers: LCCN 2017015576 (print) | LCCN 2017031028 (ebook) | ISBN 9781315375168 (ebook) | ISBN 9781138028623 (hardcover : alk. paper)
Subjects: LCSH: Rock mechanics. | Geotechnical engineering.
Classification: LCC TA706 (ebook) | LCC TA706 .S25 2017 (print) | DDC 624.1/5132—dc23
LC record available at https://lccn.loc.gov/2017015576

Table of contents

Acknowledgements

This book has been prepared on the basis of the outcomes of theoretical and experimental research works carried out by many of the graduate students as well as undergraduate students who studied at the Rock Mechanics Laboratory, Kobe University, Japan, during the past 40 years. If it had not been for former students' continuous efforts, this book would not have been published. The author extends his gratitude to all the former students for their great contributions to rock mechanics research performed at Kobe University. The author also heartily acknowledges the great support from rock mechanics research colleagues, working together on back analyses in the geotechnical engineering field. One of the chapters of this book on monitoring slope stability by using GPS in geotechnical engineering was written by Prof. N. Shimizu, Yamaguchi University, Japan. Regarding GPS displacement monitoring, its suggested method was established under the leadership of Prof. Shimizu, and it was officially approved by the ISRM Board as "ISRM Suggested Method for Monitoring Rock Displacements Using the Global Positioning System". The author would like to heartily thank him for his great contribution to Chapter 23. Thanks are also due to my wife Motoko, and daughter Junko, for their continuous encouragement and kind support during the course of preparing the manuscript.

Acknowledgements

About the author

Born in 1935, Prof. Sakurai studied Civil Engineering, first at Kobe University (B.E., 1958), then at Kyoto University (M.E., 1960), and finally at Michigan State University USA (Ph.D., 1966), having received his Dr Eng. from Nagoya University in 1975.

Prof. Sakurai worked at Kobe University, where he held the position of Associate Professor (1966–73) and Professor in the Division of Rock Mechanics, Dept. of Civil Engineering (1973–1999), and then worked at the Hiroshima Institute of Technology as President (1999–2003). He is now Professor Emeritus of Kobe University, and also Professor Emeritus of Hiroshima Institute of Technology. Prof. Sakurai worked as President of the Construction Engineering Research Institute Foundation (CERIF) (2003–2011).

In 1978–79 he was Guest Professor at the Federal Institute of Technology Zurich ETHZ, Switzerland, and in 1984 Visiting Professor at the University of Queensland, Australia. He was also Visiting Professor at Graz University of Technology in 1998.

He has given lectures in Brazil, Canada, China, Czech Republic, Germany, Greece, India, Indonesia, Italy, Kazakhstan, Korea, Poland, Taiwan, Thailand, Russia, and many other countries.

In the ISRM, Prof. Sakurai was Vice-President at Large (1988–91), President of the Commission on Communications (1987–91), and Member of the Commissions on Computer Programs (1978–87), on Rock Failure Mechanisms in Underground Openings (1981–91), and on Testing Methods (1983–91). He was also Vice-President of the Japanese Committee for ISRM (the ISRM NG JAPAN) (1995–99).

Professionally, Prof. Sakurai has been involved in various kinds of Rock Mechanics projects (hydropower, nuclear power, pumped storage and compressed air energy storage schemes; highway and railway tunnels; slopes), in Japan and abroad.

His research activities have been principally connected to numerical and analytical methods, back analysis, and field measurements, the aim of these activities being mainly concerned with making a bridge between the theory and practice. Prof. Shunsuke Sakurai is the author or co-author of over 100 publications, and the editor of "Field Measurements in Geomechanics" (Proceedings of the 2nd International Symposium, Kobe, 1987).

Prof. Sakurai received the IUE Award (1974), the JSCE Prize for the Best Paper (1990), and the ICMAG Award for Significant Paper (1994). He also received the Science Award of Hyogo Prefecture (1997).

Chapter 1

Introduction

1.1 AIMS AND SCOPE

This book is dedicated to practising engineers working in rock engineering practice, as well as for graduate students studying and doing research on rock mechanics and rock engineering. The aim of this book is to make the engineers and the students understand how to apply the theory of rock mechanics to engineering practice, in order to achieve the rational design and construction of rock structures such as tunnels, underground caverns, and slopes, and to assess not only the stability of them during/after construction, but also to ensure the safety of the workers.

In order to verify the adequacy of the original design and assess the stability of the rock structures during construction, observational methods are extremely useful. In the method field measurements play a major role, but the measurement data are only numbers unless they are properly interpreted. Back analyses can be used for interpreting the data quantitatively, resulting in the rational design and construction of the structures being achieved.

It is noted that back analysis is a highly non-linear problem, even in the simple case of linear elastic materials. This non-linearity of back analyses may attract the interest of mathematicians in back analysis problems, but only from the mathematical point of view. However, this book is not for mathematicians, but for practising rock engineers so that the back analyses should be used for engineering practice. The contents of this book are mainly based on the original works carried out in the Rock Mechanics Laboratory of Kobe University, Japan.

1.2 FIELD MEASUREMENTS AND BACK ANALYSES

Rock structures such as tunnels, underground caverns, vertical shafts, slopes, etc. are constructed with natural rocks whose geological and mechanical characteristics are extremely complex. This complexity causes difficulty in the evaluation of mechanical characteristics of rock masses, even though various laboratory and *in situ* tests, such as plate bearing tests and direct shear tests, have been developed for determining the mechanical properties of rock masses, such as Young's modulus, strength parameters, underground water condition, permeability, etc. which are important data for design analyses. In addition, the initial stresses of rock masses caused by gravitational and tectonic forces are also important data for the analyses.

It should be noted that the difficulty in the evaluation of the mechanical characteristics and initial stresses of rock masses is a characteristic feature for the design of rock structures, as the mechanical behaviours of the rock structures are extremely complex. This is entirely different from other structures like bridges and buildings, which are built with artificial materials, such as concrete and steel, whose mechanical parameters can be easily determined in laboratory experiments. Moreover, the external forces acting on the structures are also well documented.

In the mechanical behaviours of the rock structures, various uncertainties are involved not only in the mechanical characteristics of rock masses, but also in the design and construction procedures of rock structures. For instance, in tunnel engineering practice the following uncertainties are involved; (1) geological and geomechanical characteristics of rock masses are complex, (2) mechanical modelling of rock masses is extremely difficult, (3) the initial stresses of rock masses are difficult to evaluate, (4) interaction mechanism between tunnel support structures and surrounding rock masses is complex, (5) the mechanical behaviour of tunnels seems to be different for different excavation methods, (6) the mechanical behaviour is also influenced by the skill of tunnel excavation workers, etc.

In rock engineering practice, it is well known that the real behaviour of the rock structures quite often differs from that predicted by numerical analyses carried out at the design stage, even though sophisticated computer programs are used. This discrepancy may be simply because of the various uncertainties described above being involved.

In order to fill in the gap between real and predicted behaviours of rock structures, field measurements are carried out to verify the input data used in the original design, as well as to assess the stability of the rock structures during construction. In addition, it can verify the safety of the workers during the construction. Field measurements are also performed for monitoring long-term stability, for instance the monitoring of landslides. There are many different types of field measurements available, but displacement measurements are most commonly used in rock engineering practice, because they are reliable and easily handled in comparison to others such as stress and strain measurements.

However, it should be noted that the field measurement data are only numbers unless they are properly interpreted. Therefore, the most important aspect of field measurements is the quantitative interpretation of measurement results. For this purpose, back analyses must be a powerful tool.

Chapter 2

Back analysis and forward analysis

2.1 WHAT IS BACK ANALYSIS?

In back analyses, input data are measured values, such as displacements, strains, stresses and pressures, while the output results are the mechanical parameters of rock masses, such as Young's modulus, Poisson's ratio, strength parameters (cohesion and internal friction angle), permeability, and even the initial state of stress. This analysis procedure is entirely a reverse calculation of an ordinary analysis, so that it is called "back analysis", while an ordinary analysis is called "forward analysis" all through this book.

In the design of rock structures, forward analyses (ordinary analyses) are carried out for calculating stresses, strains and displacements of rock masses. The analyses require the input data which are external forces (initial stresses), the mechanical parameters of rock masses, such as Young's modulus, Poisson's ratio, strength parameters (cohesion and internal friction angle), permeability, etc. On the other hand, in the back analyses, the input data are measurement results, such as displacements, strains, stresses, etc., while the output results are the mechanical parameters of rock masses, initial stresses, permeability, etc. It is obvious that the output results of the back analyses correspond to input data of the forward analyses, while the input data for the back analyses are the measurement data. Therefore, the back analyses seem to be entirely a reverse calculation of the forward analyses, as shown in Figure 2.1.

In forward analyses, it is obvious that any sophisticated computer program can be used, no matter how many input data are needed, as long as all the data can be determined by laboratory and *in situ* tests, while in back analyses only a limited number of measurement data (input data for back analyses) are available. This means that all the input data necessary for the forward analyses are hardly identified by back analyses. To overcome this difficulty, a constitutive equation of rock masses used in back analyses should be simple enough to be able to back-calculate all the mechanical parameters of the equation from a limited number of field measurement data.

It should be emphasised that one of the important purposes of field measurements is to monitor whether the present situation of rock structures is stable, or whether an unexpected mechanical behaviour seems to start occurring. To accomplish this purpose, the field measurement results must be properly interpreted during the constructions without delay. To meet this requirement, the back analyses should be capable

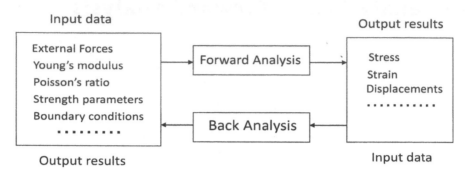

Figure 2.1 Definition of back analysis.

not only of assessing the adequacy of the original design, but also of predicting the catastrophic failure of the structures during the constructions.

2.2 DIFFERENCE BETWEEN BACK ANALYSIS AND FORWARD ANALYSIS

In forward analyses, firstly the mechanical model of rock masses is assumed as to be such as elastic, elasto-plastic, rigid-plastic, visco-elastic, etc., and the mechanical parameters of the model are determined by laboratory and *in situ* tests. Once all the mechanical parameters are determined, we can calculate displacements, strains and stresses of rocks as the outcomes of the forward analyses. This computation procedure provides a one-to-one relationship between the input data and the output results, because modelling (assumption) is done before the determination of input data, as shown in Figure 2.2. This implies that it is extremely important for the forward analyses to assume the most suitable mechanical model for rock masses.

On the other hand, in back analyses we first obtain field measurement data (displacements, strains, stresses, etc.) during constructions. These data are used as input data for back analyses, as seen in Figure 2.2. In order to perform back analyses for determining the mechanical parameters, we must assume a mechanical model. It is obvious that the mechanical parameters determined by the back analysis depend entirely on what mechanical model we assume in the back analyses. For example, if we assume an elastic model, then Young's modulus can be determined, but if a rigid-plastic model is assumed, then Young's modulus cannot be determined. Instead plastic parameters such as cohesion and internal friction angle can be obtained, though the identical input data (measurement results) are used for the both cases. This means that in the back analyses, a one-to-one relationship between the input data and output results cannot be guaranteed, because that mechanical modelling of rock masses is located in-between the input data and output results, as shown in Figure 2.2. In other words, in back analyses a one-to-one relationship between the input data and output results cannot be substantiated.

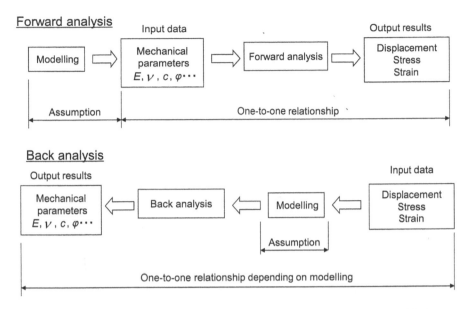

Figure 2.2 Difference between forward analysis and back analysis (Sakurai, 1997a).

We can now conclude that back analysis is not simply a reverse calculation of the forward analysis. Its concept should be different from forward analysis in such a way that back analysis should identify the mechanical model, as well as determine its mechanical parameters from field measurement results.

The mechanical model of rock masses is assumed in the design of structures, and usually only its mechanical parameters are determined by back analyses. In addition, it is noted that the back analyses determining mechanical parameters are a non-linear problem, even for the case of simple linear elastic problems, resulting in that back analyses may attract the interest of mathematicians to solve the non-linear problems. Since the mathematicians are interested in obtaining a stable solution in back analyses only from a mathematical point of view, it does not matter which mechanical model is used in back analyses.

2.3 BACK ANALYSIS PROCEDURES

2.3.1 Introduction

Back analysis problems can be solved by various approaches. Among them, inverse and direct approaches are commonly used in geotechnical engineering practice (Cividini et al., 1981). In the inverse approach the formulation is just the reverse of that in the forward stress analysis, even though the governing equations are identical. On the other hand, the direct approach is based on an iterative optimisation procedure which corrects the trial values of unknown quantities in such a way that the discrepancy between the measured and the computed quantities is minimised. In both inverse and direct methods, the number of measured values should be greater than the number of unknown quantities, otherwise the results cannot be determined uniquely.

However, it is often difficult to determine these values precisely because of the various uncertainties which are usually involved in rock engineering problems. To overcome this difficulty, a probabilistic approach is preferable as it is capable of taking these uncertainties into account. The most advantageous feature of this approach is that the final results are expressed in statistical terms, such as mean and variance.

2.3.2   Inverse approach

The inverse approach requires a mathematical formulation in a reverse way to the ordinary stress analysis so that it is only available for the linear elastic materials, whose stress-strain relationship is expressed in a linear form.

A simple example for the inverse approach in rock engineering practice is *in situ* rock tests, such as a plate bearing test, where a displacement δ is measured under a given external force P, as shown in Figure 2.3. Young's modulus E can then be determined by Equation (2.2), which is derived in a reverse formulation of the conventional stress analyses. If the number of the data (measured displacements) is greater than that of back-analysed quantities (Young's modulus, initial stresses, etc.), the least squares method can be used (Sakurai & Takeuchi, 1983).

As another example in the rock engineering field, Kovari and his colleagues (1977) developed an inverse approach called the "integrated measuring technique" for determining the rock pressure acting on tunnel linings from the strain measured on the inner surface of the lining. In this back analysis approach, mathematical equations relating the rock pressure to the axial force and bending moment of the tunnel lining were derived by imposing the equilibrium conditions between the rock pressure and the normal force and bending moment of the lining, which used a fundamental equations to determine the rock pressure acting on the lining.

For more complex engineering problems, an inverse approach can be used on the basis of a Finite Element Method (FEM), which was originally developed for structural engineering problems (Kavanagh & Clough, 1971). In the field of rock mechanics Gioda (1980) modified Kavanagh's algorithm to back-calculate both the bulk and shear moduli by applying static condensation and the least squares method.

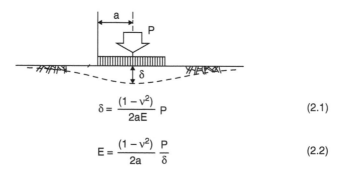

$$\delta = \frac{(1 - v^2)}{2aE} P \qquad (2.1)$$

$$E = \frac{(1 - v^2)}{2a} \frac{P}{\delta} \qquad (2.2)$$

Figure 2.3   In situ plate bearing test for determining Young's modulus from measure displacements due to applied external force.

The method is defined as "inverse", with respect to the corresponding stress analysis, since it requires the inversion of the equations governing the linear elastic stress analysis problem. If the inversion of the governing equations is possible, this technique is easily applied to engineering practice, because iteration is not necessary in computation, resulting in computation time becoming less compared with the other back analysis approaches.

2.3.3 Direct approach

The direct approach is based on the minimisation of the discrepancy between the field measurement data and the corresponding numerically evaluated quantities, in such a way that an error function δ shown in Equation (2.3) is adopted to define the discrepancy between the measured displacements and those derived from a numerical analysis.

$$\delta = \frac{\sum_{i=1}^{N} (u_i^m - u_i^c)^2}{\sum_{i=1}^{N} u_i^m} \rightarrow \min \tag{2.3}$$

where u_i^m and u_i^c are measured and computed displacements, respectively. N is number of measuring points.

The direct approach has a great advantage in avoiding the inversion of the mathematical equations of stress analyses, resulting in that it can be easily applied to any non-linear problems.

The error function defined by Equation (2.3) is in general a complicated non-linear function of the unknown quantities, and in most cases the analytical expression of its gradient cannot be determined. This is particularly evident for non-linear or elasto-plastic problems. Therefore, the function minimisation algorithm adopted for the problem solution must handle non-linear functions and it should not require the analytical evaluation of the function gradient. The algorithms meeting with these requirements, known in mathematical programming as direct search methods, are based on iterative procedures which perform the minimisation process only by successive evaluations of the error function given in Equation (2.3). Each evaluation requires a stress analysis of the geotechnical problems on the basis of the trial value chosen for the iteration.

In the minimisation algorithm for the error function, any standard algorithms of mathematical programming, such as the Simplex method (Nelder & Mead, 1965), Rosenbrock algorithm (Rosenbrock, 1960), Powell method (Powell, 1964), Conjugate Gradient method (Fletcher & Reeves, 1964), etc. can be used. However, these methods require rather time-consuming computations since a large amount of iteration is usually needed.

Gioda & Maier (1980) demonstrated the applicability of the direct method to back-calculate the non-linear material parameters and the load conditions, using a numerical example of a pressure tunnel test.

2.3.4 Probabilistic approach

Both the inverse and direct methods are based on a deterministic concept, and provide precise values for material constants and load parameters. However, it is often difficult to determine these values quantitatively because of the various uncertainties being included in rock engineering problems. To overcome this difficulty, a probabilistic approach is preferable as it is capable of taking these uncertainties into account. The most advantageous feature of this approach is that the final results are expressed in statistical terms, such as mean and variance.

The field measurement data are in general affected by various errors that depend on the nature of measured quantities, the characteristics of measuring devices, field conditions, etc. In order to evaluate the influence of these errors on the back-calculated mechanical parameters, various methods have been proposed. Among them a simulation technique, such as the Monte Carlo simulation, can be easily applied to engineering practice (Cividini & Gioda, 2003). This simulation technique is an extremely simple implementation, but requires a computational effort which rapidly increases with the increase in number of unknown parameters. In order to overcome this drawback, a probabilistic Bayesian approach is recommended (Cividini et al., 1983).

A typical feature of the Bayesian approach is that *a priori* information on the unknown parameters can be introduced in the back analysis, together with the data deriving from *in situ* measurements. In most cases, the *a priori* information consists of an estimation of the unknown parameters based on the engineer's judgement or on available general information. This leads to a numerical calibration procedure that combines the knowledge deriving from previous similar problems with the results of the *in situ* investigation.

2.3.5 Fuzzy systems, Artificial Intelligence (AI), Neural network, etc.

In a probabilistic approach, the determination of a probability density function for the mechanical parameters of rock masses is extremely difficult. In other words, there is no reliable way to determine the input data for the probabilistic approach. This is entirely different from the case of materials such as steel and concrete, resulting in that the probabilistic approach may be less applicable to rock engineering problems. To overcome this problem, the Fuzzy Set Theory can be used, which can easily provide with all the input data necessary for the back analyses on the basis of engineers' subjective judgements (Zadeh, 1965). This means that the Fuzzy Set Theory goes well with the probabilistic approach of back analyses. It should be noted that the Fuzzy Set Theory must be a potential tool for solving rock mechanics problems in probabilistic approaches (Fairhurst & Lin, 1985; Nguyen & Ashworth, 1985; Sakurai & Shimizu, 1987).

It is obvious that rock masses are extremely complex non-linear systems that include many parameters. In order to solve these complex systems, Feng et al. (2000) proposed a new displacement back analysis approach which is based on a combination of a neural network, an evolutionary technique and numerical analysis methods to identify the mechanical parameters. The method has been successfully applied to the Three Gorges Project permanent shiplock to estimate the mechanical parameters of rock masses.

Feng et al. (2004) also proposed another displacement back analysis method to identify the mechanical parameters based on hybrid intelligent methodology, such as the integration of evolutionary Support Vector Machines (SVMs), numerical analysis and a genetic algorithm.

Considering various uncertainties and complexities involved in rock masses, Khamesi et al. (2015) proposed a novel, intelligent back analysis method for determining the complex and non-linear relation between the displacements and the geomechanical parameters by using a fuzzy system designed by three different methods, i.e. nearest neighbourhood clustering and gradient descent training, particle swarm optimisation, and imperialistic algorithm.

2.4 BRIEF REVIEW OF BACK ANALYSIS

In the early 1970s, identification theories were developed in the field of system engineering (Astrom & Eykhoff, 1971), and applied to various field problems such as structural dynamics (Hart & Yao, 1977). In geomechanics, in earlier times various terms such as identification, characterisation, inverse analysis, etc., were used for identification problems. At that time it was thought that these identifications were mathematical problems, because they are highly non-linear problems, even though a simple elastic model is assumed. Therefore, the main interest of researchers has been on how to solve such non-linear problems so as to obtain a mathematically stable solution with high accuracy. Before the term "back analysis" was being used in rock mechanics field, Sakurai (1974) assumed the ground medium consisting of a visco-elastic material, and proposed a back analysis method to determine the initial stress and mechanical properties of visco-elastic underground media.

The term of "back analysis" appeared for the first time in the rock mechanics field in a paper entitled "Determination of rock mass elastic moduli by back analysis of deformation measurements" (Kirsten, 1976). Ever since that time, various names have been used by different authors. Nevertheless, the term "back analysis" gradually became popular, and it is now commonly used in the rock engineering community. In the rock engineering field, various back analysis procedures have been extensively developed, ranging from simple elastic problems to far more complex non-linear problems, and many papers related to back analysis have been published with particular reference to the interpretation of field measurement results (Gioda & Sakurai, 1987). In rock engineering practice, back analyses are nowadays often used for determining the mechanical properties of rock masses from the data of field measurements carried out during the construction.

Deterministic back analysis procedures are roughly classified into two categories: the inverse approach and the direct approach (Cividini et al., 1981). In the inverse approach, the mathematical formulation is just the reverse of that in an ordinary analysis (forward analysis in this book), although the governing equations are identical.

In the case of a ground represented by a simple mechanical model with simple geomechanical configurations, the closed-form solutions in the theory of elasticity and plasticity may be used. However, for the ground with an arbitrary shape under a more complex geological and geomechanical environment, numerical methods such as FEM, Boundary Element Method (BEM), Discrete Element Method (DEM), etc., seem to be more promising. For example, Kavanagh (1973) proposed a back analysis formulation based on FEM which may make it possible to obtain the material constants not only for isotropic materials, but also for inhomogeneous and anisotropic materials, from both measured displacements and strains.

Gioda (1980) modified Kavanagh's algorithm to back-calculate both the bulk and shear moduli by applying static condensation and the least squares method. In order to obtain the material constants, the displacements alone are sufficient. However, to identify the load conditions in addition to the material constants, the measurements of not only the displacements, but also the values for the loads and pressures are necessary. For this, a numerical procedure of back analysis was proposed for determining the earth pressure acting on tunnel lining on the basis of measured displacements and measured earth pressure at some locations. The optimal earth pressure distribution can be determined by minimising a suitably defined error function (Gioda & Jurina, 1981).

Sakurai & Takeuchi (1983) proposed an inverse method of determining both the initial stress and Young's modulus from measured displacements around a tunnel, assuming homogeneous and isotropic linear elastic media. According to the method, the strain distribution around a tunnel can be determined by the data of a limited number of measured displacements. Since the method is formulated in the stiffness matrix method, the large simultaneous equations have to be solved, resulting in time-consuming numerical computation. To overcome this shortcoming, Sakurai & Shinji (1984) used the flexibility matrix method for solving the identical problem, resulting in drastically reduced computation time. Shimizu & Sakurai (1983) extended the back analysis procedure proposed by Sakurai & Takeuchi (1983) to the three-dimensional case by using BEM to determine both Young's modulus and the *in situ* stress from measured displacements. If the back analyses are carried out with the displacements measured during the excavation of pilot tunnels for underground powerhouse caverns, the back-calculated values are those for the three-dimensional large extent of rock masses, so that they are used for assessing the adequacy of the original design of powerhouse caverns.

Gioda & Maier (1980) demonstrated the applicability of the direct method to back-calculate the non-linear material parameters together with the load conditions by introducing a numerical example of a pressure tunnel test. Cividini et al. (1985) also stated that the direct approach could be employed to determine the time-dependent material constants by applying convergence displacement measurement data taken at various stages of the tunnel construction.

Since various uncertainties are involved in rock engineering problems, it is difficult to determine the mechanical parameters of rock masses quantitatively. To overcome this difficulty, a probabilistic approach is preferable as it is capable of taking these uncertainties into account.

Among various probabilistic procedures, the Monte Carlo simulation can be easily applied to engineering practice (Cividini & Gioda, 2003). This simulation technique is an extremely simple implementation, but requires a computational effort which rapidly increases with the increase in number of unknown parameters. In order to overcome this drawback, the Bayesian approach is promising for back analyses. Cividini et al.

(1983) used Bayesian approach for the parameter estimation of a static geotechnical model.

It is obvious that one can use an ordinary probabilistic approach to solve rock engineering problems. However, the determination of a probability density function for the mechanical parameters of rock masses is extremely difficult. This means that the probabilistic approach may be less applicable to practical engineering problems. To overcome this problem, the Fuzzy Set Theory can be used, which can easily provide with all the input data necessary for the back analyses on the basis of engineers' subjective judgements. This means that the Fuzzy Set Theory goes well with the probabilistic approach of back analyses. It should be noted that the Fuzzy Set Theory must be a potential tool for solving rock mechanics problems in probabilistic approaches (Fairhurst & Lin, 1985; Nguyen & Ashworth, 1985; Sakurai & Shimizu, 1987).

As an alternative probabilistic approach, the Kalman filter theory (Kalman, 1960) must be a potential approach for back analyses in Geomechanics. Murakami & Hasegawa (1985) proposed to use the Kalman filter theory for applying geotechnical problems.

The back analysis of the *in situ* stress in a non-linear material was formulated by Zhang et al. (1988) using an iterative back analysis algorithm based on BEM. Yang & Sterling (1989) also proposed a unique back analysis method for determining both the *in situ* stress and the elastic properties of rock masses using the fictitious stress boundary element method.

In any case of the back analyses described above, the mechanical models are assumed to be the same as those used in the design analysis. However, Sakurai (1997a) warned that in back analyses a mechanical model should not be assumed, but should be identified by back analyses. This is due to the fact that in forward analyses a one-to-one relationship between input data and output results is guaranteed, while in back analyses the output results depend entirely on what mechanical model is assumed, as shown in Figure 2.2.

To overcome this problem, Sakurai et al. (1993b) proposed a back analysis method considering "non-elastic strain", which can be applied to represent any type of non-linear mechanical behaviours of geomaterials without assuming any mechanical model. This method is known as a "universal back analysis method", as described in Chapter 9. The applicability of the method is demonstrated by showing a case study of a shallow tunnel, which reveals that the method can well simulate non-linear behaviours, such as a loosening zone occurring above the tunnel crown arch (Sakurai, 1996).

Considering the difficulty in developing the constitutive equations, because of many uncertainties being involved in mechanical characteristics of geomaterials, Feng & Yang (2004) proposed a hybrid evolutionary algorithm for coupling recognition of the structure of the non-linear constitutive material model and its coefficient in global space using global response information, such as load versus deflection data, obtained from the structural tests. Genetic programming is used to recognise the structure of the non-linear stress-strain relationship without making any assumption in advance, and the genetic algorithm is then used to recognise its coefficients.

The response of rocks to stress is highly non-linear, so it is difficult to establish a suitable constitutive model using traditional mechanics methods. In order to provide adequate models for the time-dependent behaviour of rocks, Feng et al. (2006) proposed a new identification method based on hybrid genetic programming with the improved Particle Swarm Optimisation (PSO) algorithm, which was used to identify the visco-elastic models for rocks.

Khamesi et al. (2015) proposed a novel, intelligent back analysis method combining fuzzy systems, imperialistic competitive algorithm and numerical analysis.

Feng & Hudson (2010) discussed how to establish the necessary quality and quantity of information required for modelling and designing in rock engineering. In rock engineering projects, observational methods are the most suitable methodology where design, construction and monitoring must be closely related to each other. It is obvious that back analysis plays an important role in these observational methods. It is also important to feed the information obtained by back analyses into the design and construction of the structures without delay. In the design of structures, we can use any sophisticated computation scheme, while in the monitoring of structures a back analysis technique should be simple enough to be able to feed the results back into engineering practice.

Moreover, it is obvious that back analysis is the complete opposite to forward analysis in terms of input data and output results, as seen in Figure 2.1. This means that it may be possible to formulate constitutive equations only for the use of back analyses, on the basis of a completely different conception from forward analyses. Based on the different conception for back analysis, Sakurai & Shinji (2005) proposed a simple constitutive equation in which the anisotropic parameter (anisotropic damage parameter) plays an important role, and it is well applicable to the back analyses of tunnels and slopes without assuming any mechanical model. Moreover, in the proposed constitutive equation, it is not necessary to assume any type of mechanical modes, i.e., strain hardening, perfectly plastic or strain softening, which can be automatically determined by back analyses of measured displacements. (See Section 14.5).

In the design of slopes the factor of safety defined by the ratio of the shear strength of geomaterials to sliding force acting along a sliding plane (as given in Equation (17.1)) is commonly used, while in monitoring the slope stability, displacement measurements are usually carried out, resulting in no factor of safety being assessed during constructions. This is a paradox between the design and monitoring of slopes. To overcome the paradox, Sakurai & Nakayama (1999) proposed a back analysis procedure to determine the strength parameters, such as cohesion and internal friction angle of geomaterials from measured displacements, so that the factor of safety can be determined during constructions. In the back analysis procedure, the critical shear strain of geomaterials defined by the ratio of the maximum shear strength of geomaterials to their shear modulus (as shown in Equation (7.1)) plays a major role, making it possible to link the monitoring to the design of slopes. Sakurai et al. (2009) applied the proposed back analysis procedure to assess the slope stability of an open-pit mine during excavations, resulting in that the back-calculated factor of safety became approximately $FS = 1.0$ in one day before a failure occurred. This fact demonstrates that the proposed back analysis procedure is well applicable to slope engineering practice.

Chapter 3

Modelling of rock masses in back analysis

3.1 MODELLING OF ROCK MASSES

Rock masses are generally classified into three groups (a) continuous, (b) discontinuous, and (c) pseudo-continuous types, as shown in Figure 3.1.

Type (a) represents the rock masses consisting of rocks without any joints or discontinuities. It may also consist of soft rocks in which the strength of rock matrix is usually less than that of joints. Therefore, in Type (a) the effects of joints and discontinuities are not necessarily taken into account, even though joints and other types of discontinuities exist.

Type (b) represents jointed rock masses that are divided into a low number of blocks by joints and discontinuities. In reality, however, rocks have many different kinds of joints and discontinuities and it is almost impossible to model all the discontinuity systems. In engineering practice, therefore, only dominant discontinuities are taken into account. This type of rock can be analysed by discontinuous approaches proposed by Goodman et al. (1968), Cundall (1971), Kawai (1980) and others.

Type (c) is for highly fractured and/or weathered rock masses that may exhibit an overall behaviour being similar to that of a continuous body. Therefore, it is called a pseudo-continuous type.

The difference between Type (b) and Type (c) is the size of structures built in or on the ground. The rock mass is classified into Type (c) when the size of structures is quite large compared with the spacing of joints and discontinuities. When the structure is small, however, the identical rock is classified into Type (b), as shown in Figure 3.2.

It is a specific feature of rocks such that the mechanical characteristics change with the size of structures. Therefore the mechanical constants, which are used as input data in the computational analyses, must be properly evaluated with consideration to the size of the structure. This implies that even the mechanical constants determined by *in situ* tests, such as plate bearing tests and direct shear tests, cannot be used directly for analysing the behaviour of structures, unless they are extrapolated to obtain the values for rocks in a larger extent comparable to the size of structures, even though this extrapolation is not an easy task. In order to overcome these difficulties, field measurements are carried out during constructions, so that the mechanical constants such as Young's modulus, Poisson's ratio, cohesion, internal friction angle, etc. are back-calculated from the field measurement results. The initial stresses of rock masses are also back-calculated from field measurement results.

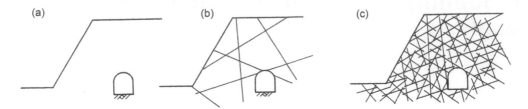

Figure 3.1 Classification of the ground: (a) continuous type, (b) discontinuous type, and (c) pseudo-continuous type.

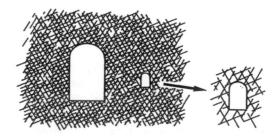

Figure 3.2   Rock mass classification depending on the size of structures.

In back analyses procedures described in this book, both Type (a) and Type (c) are considered, while Type (b) must be considered by the discontinuous approach, which is excluded in this book.

3.2   BACK ANALYSIS AND MODELLING

In the numerical analyses, the mechanical modelling of rock masses is not an easy task, because geological and geomechanical characteristics of rock masses are extremely complex. In forward analyses, the mechanical model of rock masses is usually assumed as a certain model such as elastic, elasto-plastic, visco-elastic-plastic, discrete block models, etc. The mechanical constants of the materials can be determined by laboratory and/or *in situ* tests. Once all the mechanical constants are determined, we can calculate displacements, strains and stresses occurring in the rock masses. The results of the calculation give exact values at least under an assumed mechanical model with the mechanical constants determined by laboratory and/or *in situ* tests, resulting in that uniqueness of the solutions is substantiated between the input data and output results, as shown in Figure 2.2.

In back analyses, on the other hand, field measurements are carried out to measure displacements, strains, and stresses occurring in rock masses, and then the back analyses are performed to determine the mechanical constants of the materials from the measurement results by assuming a mechanical model. It is no wonder that the values of mechanical constants determined by the back analyses depend entirely on what model we assume in back analyses. For instance, if we assume an isotropic linear elastic model, then Young's modulus can be obtained, but if an elasto-plastic model is assumed, then Young's modulus and plasticity constants such as cohesion and internal friction angle can be back-calculated, even though the identical values of measurement results are used as input data. This means that the results of back analyses are basically a matter of assumption of the mechanical models representing the mechanical behaviour of geomaterials. In other words, in back analyses the uniqueness of solutions cannot be substantiated between the input data and output results, as shown in Figure 2.2. This means that the back analyses in geotechnical engineering practice should be capable of identifying not only the mechanical constants, but also the mechanical models.

Regarding the mechanical modelling of geomaterials, there is another problem; when a back analysis procedure is newly developed, it is important to verify its applicability in the accuracy and reliability of the procedure. For this purpose, the applicability of the newly developed procedure is demonstrated by showing an example problem, where a forward analysis is firstly carried out to determine displacements, strains and stresses under given input data. After that, the calculated displacements are used as input data for the newly developed back analysis procedure, resulting in obtaining the outcomes, i.e., mechanical parameters. If the mechanical parameters determined by the back analysis are the same as those used as the input data in the forward analysis, the accuracy and reliability of the newly developed back analysis procedure will have been verified. In this demonstration, the identical mechanical model is used for both forward and back analyses, because this is only the demonstration of the applicability of the newly developed procedure, so that it does not matter what mechanical models are used.

Furthermore, in rock engineering practice, one of the important purposes of back analyses must be to predict catastrophic failures (unexpected failures) of the structures from field measurement results during the constructions. To achieve this purpose, in the back analyses the mechanical models should not be assumed, but it should be back-calculated from the measurement results, because the unexpected failure modes are usually not taken into account in the mechanical models used in design analyses.

3.3 DIFFERENCE BETWEEN PARAMETER IDENTIFICATION AND BACK ANALYSIS

Parameter identifications and back analyses must be different from each other. In the parameter identifications, the mechanical constants of rocks used in the design analyses are verified by field measurement results during the constructions, and if necessary they would be modified, while the mechanical model remains unchanged all the time. Therefore, it should be noted that the parameter identifications can be applied only when the mechanical models are well defined and fixed.

However, the mechanical characteristics of geomaterials such as soils and rocks are so complex that it is extremely difficult to define the mechanical model to represent their mechanical behaviour. In fact, many studies have been carried out, but there still remains many problems for the modelling of geomaterials. This means that the identification of mechanical parameters alone is insufficient, but the mechanical model of rock masses must be identified by the field measurement results. To overcome the difficulties in modelling of mechanical characteristics of rock masses, the mechanical

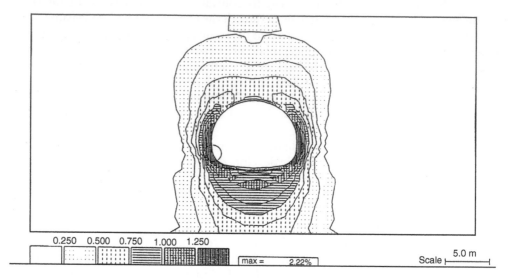

Figure 3.3 Back analysis result (isotropic elastic model being assumed) (Sakurai, 1997a).

model of rock masses should be identified by the field measurements carried out during constructions. The identification of mechanical model together with the mechanical constants of rock masses must be an important aspect of back analyses. In other words, the back analyses should make it identify both the mechanical model and mechanical constants.

The parameter identifications determining the mechanical constants should be distinguished from the back analyses, in which both the mechanical parameters and mechanical model are identified. This identification must be defined as a back analysis, which can make it determine both the mechanical model and mechanical constants of rock masses from measured quantities. In conclusion, the mechanical model of rock masses should not be assumed, but it should be identified by back analyses of field measurement results.

If the mechanical model is fixed all the time, the outcomes of back analyses are not only inadequate, but also provide misleading results for interpreting the field measurement results. As a result, back analysis results provide wrong information for assessing the design and construction procedures.

The following example demonstrates how misleading results may be derived if we assume an inadequate mechanical model for back analyses in tunnel engineering practice.

The tunnel concerned is for a double-track railway which is excavated in soil formation with shallow depth. The overburden height is about 10 m. The support structures are shotcrete and steel ribs. Displacements of the ground around the tunnel were measured by both extensometers and inclinometers during tunnel excavations. Surface settlements were also measured by conventional surveying of levelling.

Back analyses were carried out during the excavation in order to assess the stability of the tunnels. In back analyses two different approaches were used. One assumes

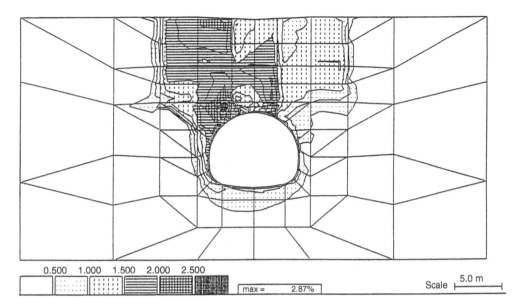

Figure 3.4 Back analysis result (non-elastic strain being considered) (Sakurai, 1997a).

isotropic linear elastic materials (Sakurai & Takeuchi, 1983), and the other uses the "universal back analysis method" where non-elastic strain is back-calculated, resulting in that the mechanical model is not necessarily assumed (Sakurai et al., 1993b).

The maximum shear strain distributions around the tunnel obtained by back analyses are shown in Figure 3.3 and Figure 3.4. The former was obtained by assuming the isotropic linear elastic model, while the latter was determined by using the universal back analysis method (Sakurai, 1996).

It is of interest to know that these two Figures show entirely different maximum shear strain distributions, even though the identical input data (measured displacements) were used, but using different mechanical models. The result shown in Figure 3.3 is one determined by assuming an isotropic linear elastic material, while the result shown in Figure 3.4 was obtained by the "universal back analysis method" where the non-elastic strains are back-calculated. In other words, any mechanical model is assumed. The detail of the "universal back analysis method" is described in Chapter 9.

It is obvious from Figure 3.4 that a loosening zone is likely to occur above the tunnel crown arch. It is noted that this example problem demonstrates that the mechanical model in back analyses should not be assumed, but should be identified by back analyses; otherwise a back analysis may lead to a misleading conclusion in engineering practice.

Chapter 4

Observational method

4.1 WHAT IS OBSERVATIONAL METHOD?

The mechanical behaviours of rock structures such as tunnels and slopes are extremely complex, resulting that their real behaviours often differ from the ones predicted at the design stages. This discrepancy is simply because of the fact that many uncertainties are involved in both geological and geomechanical characteristics of rock masses, as well as in the initial stresses acting in them. To fill in the gap between the real and predicted behaviours, Terzaghi & Peck (1948) proposed an "observational procedure", which is a construction method for the rock structures. According to the method, the stability of the structures is monitored by the field measurements, usually displacement measurement during constructions. In addition to that, the validity of the original design can be verified, and if necessary the original design can be modified, for instance, by installing additional support countermeasures for stabilising the structures.

In the observational methods, field measurements play a major role, but as already described, the field measurement data are only numbers unless they are properly interpreted. This implies that the most important part of the observational methods is in how to interpret the field measurement results. In order to interpret the measurement results, we need the failure criterion for displacements. However, the failure criterion of rocks is usually given in terms of stress, resulting in that empirical criteria are commonly used. It is noted that empirical criteria are developed on the basis of engineer's experiences, which is called engineering judgements.

For instance, in slopes stability problems, the following empirical criteria are often used, that is, the certain values of total displacements together with the increasing displacement rates are determined as a criterion prior to constructions for monitoring the stability of slopes. These empirical criteria are popularly used in slope engineering practice, but there is no rock mechanics theory behind them.

In order to give a theoretical background for the failure criteria of rock masses, Sakurai (1981) proposed the critical strain of rock masses on the bases of the laboratory and *in situ* tests. The critical strain is described in Chapter 5.

4.2 DESIGN PARAMETERS FOR DIFFERENT TYPES OF STRUCTURES

In the design of rock structures, numerical analyses have been popularly performed. However, the accuracy and reliability of their results entirely depend on the input

Table 4.1   Difficulty in the determination of design parameters for different types of structures; (Mechanical properties of materials and external forces).

Type of structures	Mechanical properties	External forces
Bridges	easy	easy
Dam foundations	difficult	easy
Slopes	difficult	moderate
Tunnels and caverns	extremely difficult	extremely difficult

data which are mechanical parameters of rock masses, as well as the initial stresses acting in rock masses. The input data (design parameters) for the numerical analyses can be roughly classified into two groups, i.e. one is external forces, and the other is mechanical properties of rock masses. It is obvious that the accuracy and reliability of both the external forces and mechanical properties of materials depend on the types of structures. Let us discuss the accuracy and reliability of the design parameters for the following different types of structures, that is, (1) bridges, (2) dam foundations, (3) slopes, and (4) tunnels and underground caverns.

For bridges, the external forces (loads) are usually given by design codes, depending on the type of bridges, such as railway bridges, highway bridges, pedestrian bridges, etc. For dam foundations, the external forces are mainly caused by hydraulic pressure from reservoirs and the weight of the dam, which can be determined with certain accuracy. For slopes, the external forces may be mainly due to the gravitational force, but in some cases, tectonic pressures may become important, resulting in that the external force must be determined with a little caution. Lastly, for tunnels and underground caverns, the external forces must be both the gravitational force and tectonic pressures. In the numerical analyses, they are usually replaced by the so-called initial stresses of rock masses. However, it is not easy to replace the external forces by the initial stresses, because rock masses are non-homogeneous materials containing various types of discontinuities ranging from small cracks to large fractures and faults. This non-homogeneity causes the initial stresses to vary from place to place, making it extremely difficult to determine the initial stresses (external forces).

On the other hand, concerning the mechanical properties of materials, it is easy to determine those of bridges because they are usually built with concrete and steel, whose mechanical properties can be easily obtained by laboratory experiments. However, rock structures such as dam foundations, slopes, and tunnels and underground caverns, consist of natural rocks whose mechanical properties are extremely difficult to determine.

All the results of the above discussions are summarised in Table 4.1.

Looking at this table, we can see that the rock structures, particularly tunnels and underground caverns, are entirely different from bridges in terms of the design parameters. That is, for bridges the design parameters of both mechanical properties and external forces are easily determined, while for tunnels and underground caverns, the design parameters of both material properties and external forces are extremely difficult to determine.

In the design of bridges, the stresses acting in the member of the structures are calculated under the application of external forces (loads) with consideration of mechanical properties. This calculation can provide reliable results, because all the input data can

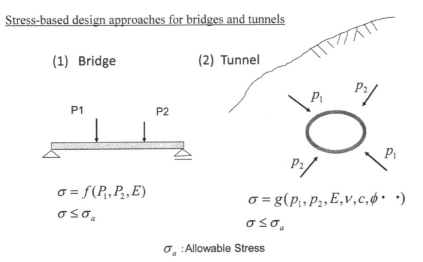

Stress-based design approaches for bridges and tunnels

(1) Bridge   (2) Tunnel

$$\sigma = f(P_1, P_2, E)$$

$$\sigma \leq \sigma_a$$

$$\sigma = g(p_1, p_2, E, v, c, \phi \cdots)$$

$$\sigma \leq \sigma_a$$

σ_a : Allowable Stress

Figure 4.1 Stress-based design approaches for bridges and tunnels.

be evaluated with high accuracy. Then, the structural design is performed so as to satisfy the requirement that the calculated stress is always smaller than the allowable stress of materials, which can also be determined easily by laboratory experiments. This design approach is commonly used not only for bridges, but also for various types of structures built with artificial materials like concrete, steel, plastic material, etc.

On the other hand however, for rock structures, particularly tunnels and underground caverns, the stress-based approach is hardly applicable for the design, because it is difficult to determine the input data, both external forces and mechanical properties. This difficulty in determining the input data causes a problem that the accuracy of calculated stresses of the structures is questionable even though we use a sophisticated computer program. In addition, the allowable stress of rock masses is also hardly determined either in laboratory experiments, or in *in situ* tests, because of the scale effect of rock masses. This results in that the comparison of calculated stresses in rocks with the allowable stress has no meaning (see Figure 4.1). In other words, the stress-based design approach cannot be applied to the design of tunnels and underground caverns. This means that the design approach for tunnels and underground caverns excavated in natural rock masses should be different from that of bridges built with artificial materials like concrete and steel.

4.3 DIFFERENCE BETWEEN STRESS-BASED APPROACH AND STRAIN-BASED APPROACH

In the stress-based approach, the factor of safety is defined as follows:

$$FS = \frac{\sigma_a}{\sigma} \tag{4.1}$$

where σ_a is the allowable stress, and σ is stress occurring in the materials.

The factor of safety defined by Equation (4.1) can be used commonly for the design of bridges and buildings. However, we have never heard of the factor of safety of tunnels. This may be due to the fact that the factor of safety of tunnels cannot be defined in terms of stress, because of the reasons described as follows.

When a tunnel is excavated in an elastic ground, its factor of safety is calculated in terms of stress, and it should be greater than FS = 1.0, as shown in Equation (4.2).

$$FS = \frac{\sigma_c}{\sigma_e} \geq 1.0 \tag{4.2}$$

where σ_c is the yielding stress of rocks, and σ_e is stress occurring in the ground.

However, when a tunnel is excavated in an elasto-plastic ground, where a plastic region appears around the tunnel, then the stress occurring in the plastic region always satisfies the yielding criterion, resulting in the stress in the plastic region being the same as the yield stress of rocks. Thus, if we define the factor of safety of tunnels as the ratio of the yielding stress to the stress occurring in the plastic region, it is obvious that the factor of safety always becomes FS = 1.0, as shown in Equation (4.3).

$$FS = \frac{\sigma_c}{\sigma_p} = 1.0 \tag{4.3}$$

where σ_p is stress occurring in plastic regions.

This indicates that the plastic regions seem to always be in a critical condition. In reality however, the tunnel does not fail even after the stress reaches the yield stress, because the plastic regions are supported by the surrounding elastic region. This may be the reason why no tunnel engineers have ever been concerned with the factor of safety of tunnels, while bridge engineers always use the factor of safety for assessing the stability of the structures. To overcome the difficulties in evaluating the factor of safety of tunnels, a strain-based approach may be advantageous, where strain occurring around the tunnels is compared with its allowable strain.

In tunnels, the strain-based approach is surely far superior to the stress-based approach. However, the superiority of the strain-based approach for assessing the tunnel stabilities is not popularly recognised among tunnel engineers. This may be due to the fact that almost all structures such as bridges, buildings, cars, aircrafts, etc. are designed by the stress-based approach, in such a way that the stresses occurring in the members of structures always remain within the elastic limit of materials. It should be noted that tunnels are the only structures which allow the materials to be used in a plastic state.

The factor of safety of tunnels for both stress-based approach and strain-based approach is summarised in Figure 4.2. It is obvious from the Figure that the strain-based approach must be superior to the stress-based approach in tunnel engineering practice.

In the design of slopes, the factor of safety is usually defined as the ratio of resistance forces to sliding forces along a sliding plane (see Equation (17.1)). The resistance forced is calculated in terms of shear strength of rocks, while the sliding forces are determined by shear stresses occurring in rocks. In the design of slopes, no displacements are taken into account. On the other hand, in monitoring, displacement measurements

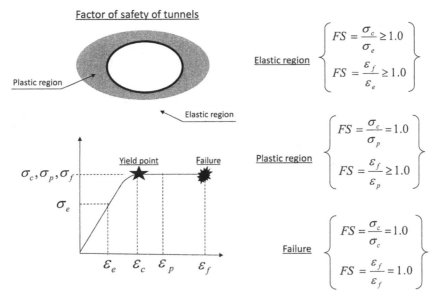

Figure 4.2 Factor of safety of tunnels (stress-based approach and strain-based approach).

are usually carried out, resulting in that the stress-based factor of safety used at the design stages cannot be assessed by measured displacements. This is a paradox in the design and monitoring of slopes (see Section 17.2), that is, stresses are considered in the design, while displacements (strains) are measured in monitoring. The question is how we can assess the stability of slopes by the measured displacements.

4.4 STRAIN-BASED APPROACH FOR ASSESSING THE STABILITY OF TUNNELS

When constructing tunnels and caverns in complex geological conditions, field measurements are usually carried out, not only for monitoring the stability of surrounding rocks, but also for assessing the adequacy of support structures such as shotcrete, rock bolts, and steel ribs. Both the original design for the support structures and the construction procedures are then evaluated, considering the results of the field measurements. They are modified if necessary. This construction method is called the observational procedure.

In order to use the observational procedure successfully in the construction of tunnels, the interpretation of field measurement results is extremely important, and its results must be properly used without delay for assessing the support measures and construction procedures. For interpreting the field measurement results properly, Sakurai (1981) proposed a strain-based approach which is called Direct Strain Evaluation Technique (DSET). The basic concept of the DSET is to assess the stability of tunnels by comparing the strain occurring in the ground surrounding the tunnels with the allowable strain of the geomaterials.

For the allowable strain of rock masses, Sakurai (1981) proposed "critical strain" in compression, which is defined as the ratio of the uniaxial compressive strength to Young's modulus of geomaterials. This is a dimensionless quantity. Based on the critical strain, "hazard warning levels" are proposed (see Section 5.4), which are classified into three levels depending on the magnitude of strains occurring in geomaterials. Following the same idea as the critical strain, Sakurai et al. (1993a) proposed "critical shear strain" of geomaterials, and derived hazard warning levels in terms of maximum shear strain, which can be used for assessing the stability of tunnels and slopes in more general cases (Section 7.2 and Section 21.2). The hazard warning levels are described more in Section 4.8.

Nevertheless, the crucial issue of the DSET is how to determine the strain distribution around the rock structures from measured displacements. It is obvious that the strains can be determined from measured displacements by using the kinematic relationship between strains and displacements. This approach is theoretically possible, but practically hard to apply, because it requires many displacement measurements to determine the strain distributions (Sakurai, 1981). To overcome this difficulty, back analyses can be used for determining the strain distributions around the rock structures from measured displacements, resulting in that the stability of structures are quantitatively assessed by the field measurement results.

4.5 DISPLACEMENT MEASUREMENTS IN OBSERVATIONAL METHOD

In the observational methods, field measurements play a major role in monitoring the stability of rock structures, such as tunnels, underground caverns, cut slopes, natural slopes, vertical shafts, etc. The results of field measurements can be used for assessing not only the stability of the structures during/after their constructions, but also re-evaluating the mechanical parameters of rock masses used in the design analyses, and if necessary, the original design is modified. In addition, the construction procedures are also modified on the basis of the field measurement results. However, in order to assess the stability of the structures, as well as to assess the adequacy of the original design and construction procedures, the measurement results must be properly interpreted. For this purpose, back analyses are very effective.

Concerning field measurements, various types of techniques with advanced equipment have been developed. Among the field measurements, displacement measurements may be most popularly adopted because they are reliable and easily handled in engineering practice in comparison with stress and strain measurements. In the displacement measurements in geotechnical engineering practice, convergence meters, extensometers (borehole extensometers), inclinometers (borehole inclinometers), total stations, GPS, GNSS, etc. are used.

In the monitoring of tunnels being excavated by the convergence-confinement methods, the crown settlements and convergences at the inner surface of tunnels are in general measured during excavations, where the measuring points should be taken as close to the tunnel face as possible. When the excavation begins, the crown settlements

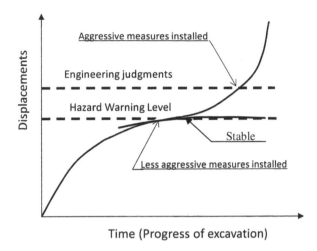

Figure 4.3 The optimal timing of installation of support measures for stabilising the structures.

and convergences are measured with the progress of excavation. If the rock masses have time-dependent mechanical characteristics (creep), the measured displacements increase with time even though there is no progress of the excavation.

For slopes, extensometers and inclinometers are commonly used for displacement measurements. Recently, GPS displacement measurements have become popular because the three-dimensional displacement components (i.e. vector components) can be obtained, resulting in that the measurement results can be used as input data for back analyses determining the mechanical parameters of geomaterials.

In any case, the displacements usually increase with the progress of excavation, but displacement rates tend to gradually become small with time. If however, the displacements start accelerating, some support measures should be installed to stabilise the structures. It is noted that in this case aggressive support measures must be needed to stabilise them. On the other hand, if additional support measures are installed before the displacements start accelerating, we can stabilise the structure with a less aggressive support measure, as shown in Figure 4.3.

To make a decision on the optimal timing when the additional support measures are installed, the criterion for displacements is needed, which is usually determined by engineering judgements, but it must be determined on the basis of rock mechanics theory. Considering this condition for the criterion, Sakurai (1997a) proposed the "hazard warning level", which is based on the critical strain of rocks and rock masses (see Chapter 5), and is extremely useful in the observational methods.

It is noted that in the observational methods the field measurement results should be interpreted correctly, and proper actions must be taken without delay. In the interpretation of the measurement results, back analyses are an effective tool for making a decision as to what type of support measures is adequate for stabilising the structures.

4.6 BACK ANALYSIS IN OBSERVATIONAL METHOD

In the observational method the mechanical parameters adopted in the original design of structures are verified by back analyses of field measurement results. In addition, the mechanical model of the geomaterials adopted in the design analyses should also be verified by field measurement results. As a result, if necessary the original design of the structures is modified. In this feedback process of the observational methods, both field measurements and back analyses play an important role to achieve the rational design of structures.

It is common that in the observational methods, mechanical models are usually assumed to be the same as that adopted in design analyses, and only their mechanical parameters are re-evaluated by back analysis of field measurement results. However, as already discussed in Section 3.3, it is of importance that the mechanical model should not be assumed, but it should be identified by the back analyses of field measurement results, or otherwise the back analysis results may derive misleading conclusions.

Now, a question may arise how to identify the mechanical model of geomaterials from field measurement results. One of the answers is to use a so-called trial-and-error method, in which several mechanical models, including one adopted in the design analyses, are firstly assumed by considering the results of both laboratory and field explorations, such as geological and geomechanical investigations, followed by back analyses (direct approach) being carried out in selecting one of the assumed mechanical models. Their mechanical parameters can then be determined from the measurement results, in such a way that the error function defined in Equation (2.3) becomes minimised. This calculation procedure is repeated by changing the mechanical models one by one, and the values of the error function for each mechanical model are compared with each other, resulting that the mechanical model which yields the smallest value of the error function must be the optimal mechanical model.

However, the trial-and-error method is questionable, because of the fact that there is no guarantee whether the mechanical model back-calculated by this method is the most appropriate model. There may be other possibilities for the mechanical models to provide further smaller values of the error function. In other words, the mechanical models which can be determined by the trial-and-error method are not necessarily the optimal mechanical models.

To overcome this shortcoming for determining the mechanical models, an alternative approach has been proposed in which the "anisotropic parameter", defined as the ratio of the shear modulus to Young's modulus, is introduced (Sakurai & Shinji, 2005). The anisotropic parameters can be determined by back analyses (direct approach) of measured displacements, in such a way that the error function defined by Equation (2.3) is minimised by changing the anisotropic parameters. When the optimal anisotropic parameters are determined, the displacements and strains occurring around the structures can be calculated by forward analyses using the optimal anisotropic parameters as input data, resulting in the stability of the structures being assessed by comparing the strains with the hazard warning levels (see Section 4.8). If the occurring strains tend to exceed the hazard warning levels, then additional support measures are installed to stabilise the structures. In this approach both the mechanical models and their mechanical parameters are not required.

It is noted that one of the important purposes of the observational methods is to monitor whether the present condition of the structures is stable, or if some unexpected failures (catastrophic failures) seem to start occurring around the structures. However, it is extremely difficult to predict the unexpected failures of structures by back analyses of field measurement results, because of the fact that in general the unexpected failures are not taken into account in the mechanical models. To overcome this problem, the proposed approach is advantageous in predicting any types of failures including unexpected failures without assuming any mechanical models, in a way that the stabilities of structures are assessed simply by comparing the strains with the hazard warning levels, which are expressed in terms of the failure strains of geomaterials.

In the observational procedures, it is noted that when the strains occurring around structures tend to exceed the hazard warning levels, the original design of the structures must be modified by installing the proper support measures. In the design of the additional support measures, the forward analyses are carried out to make it reduce the strains by changing the values of the anisotropic parameters so as to remain the strains within the hazard warning levels.

4.7   FLOWCHART OF OBSERVATIONAL METHODS

The observational method for the design and construction of rock structures is described in the flow chart shown in Figure 4.4. As seen in the Figure, in rock engineering projects various laboratory and field explorations are first carried out to collect data, such as the geological conditions, the geomechanical characteristics of the rocks, the groundwater behaviour, etc.

The collected data are used as input data for the design of structures. When the designs are complete, constructions start. During the constructions, field measurements such as displacement measurements, etc. are carried out to monitor the stability of the structures. If the measured displacements are still within the allowable values (hazard warning levels), then the structures are still stable, resulting in constructions continuing. If, however, the measured displacements reach a certain level of the hazard warning levels, back analyses are carried out to assess the validity of original designs, not only for the design parameters, but also for mechanical modelling. The back analyses considering the anisotropic parameters may be performed (see Section 4.6).

If the instability of structures is likely to start occurring, the original design and the construction procedures are modified, and additional support measures are installed so as to increase the stability of the structures, followed by the constructions proceeding until the end of the constructions, as shown in Figure 4.4.

4.8   HAZARD WARNING LEVELS

4.8.1   Introduction

As already described, the field measurement results are only numbers, unless they are properly interpreted. For interpreting the measurement results properly, the hazard warning levels for the measurement quantities are needed for assessing the stability of

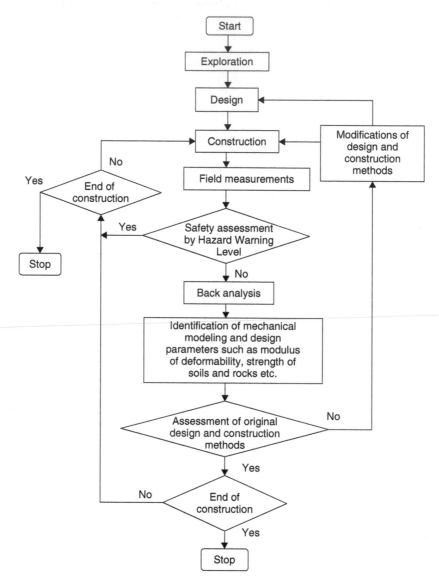

Figure 4.4 Observational methods for the design and construction of rock structures (Sakurai, 1997a).

the structures. In monitoring of the stability of structures, displacement measurements are usually performed, because of the easy and reliable data being obtained. However, it should be noted that the failure of geomaterials occurs by strains, not by displacements. Therefore, the measured displacements cannot be directly used for assessing the stability of structures, unless they are properly interpreted.

In the monitoring of structures, it is recommended that the hazard warning levels should be fixed for each measurement item prior to the start of constructions. This will make it possible to assess the stability of structures immediately after taking measurement data by comparing the measured values with the hazard warning levels. Once the constructions start, field measurements, usually displacement measurements, are performed. If the measured displacements remain within the hazard warning levels, the stability of the structures is confirmed. If the measured values tend to become greater than the hazard warning levels, the original design must be modified to stabilise the structures.

For determining the hazard warning levels, there are two approaches available; one is an empirical approach, and the other is theoretical. The empirical approaches are based on the engineer's experiences, in that the certain magnitude of displacements and/or increasing rate of displacements are often adopted as the criteria (hazard warning levels) for monitoring the stability of structures. The empirical approaches are popularly used for assessing the stability of slopes, even though there is no theoretical background for them.

On the other hand, the theoretical approaches for determining the hazard warning levels are classified into two ways; one is based on numerical analyses, and the other is based on the critical strain of geomaterials.

4.8.2 Numerical analysis methods

Since rock masses are non-homogeneous materials, their mechanical properties vary from place to place with a large scattering, because of the complexity of their geological and geomechanical characteristics. Therefore, design engineers usually carry out a conservative design in such a way that the smallest values of mechanical parameters, such as Young's modulus and strength parameters (cohesion and internal friction angle) are chosen among the scattering results as input data for the design analyses, resulting in the largest values of displacements and strains being obtained. However, in monitoring the stability of structures, it is obvious that the largest value of displacements and strains cannot be used as the hazard warning levels, because the largest values of displacements and strains lead to the dangerous side of judgements in interpreting measured displacements. Therefore, in determining the hazard warning levels by numerical analysis methods, we must choose the largest values of both Young's modulus and strength parameters as the input data for numerical analyses, resulting in the displacements and strains being smallest, which provide conservative judgements (Sakurai, 1997c).

It is noted that when we determine the hazard warning levels by numerical analysis methods, their input data must be different from the ones used in design analyses, as shown in Table 4.2. Moreover, the hazard warning levels determined by the numerical analyses are not necessarily the most suitable hazard warning levels, because the numerical analysis results entirely depend on what mechanical model is assumed.

4.8.3 Critical strain methods

To overcome the difficulties in assuming mechanical models, the critical strain of geomaterials is proposed (Sakurai, 1981), which is defined as the ratio of uniaxial compressive strength to Young's modulus, resulting in that the effects of joints existing in rock masses cancel each other out, even though both the uniaxial compressive

Table 4.2   Relation between the mechanical parameters in design and monitoring (Sakurai, 1997c).

	Design	*Monitoring*
Young's modulus		
$E_s \sim E_l$	E_s	E_l
Strength		
$q_{us} \sim q_{ul}$	q_{us}	q_{ul}
Cohesion		
$c_s \sim c_l$	c_s	c_l
Friction angle		
$\phi_s \sim \phi_l$	ϕ_s	ϕ_l
Displacements	u_l	u_s

s: small value
l: large value

strength and Young's modulus are greatly influenced by the existence of joints. In fact, the critical strain of jointed rock masses is almost identical, or slightly increases with decreasing uniaxial compressive strength, to that of intact rock specimens. This means that we can use the critical strain determined by laboratory tests on intact rock specimens for assessing the stability of a large-scale rock masses. It is also advantageous for critical strain that it is not much influenced by various aspects of the environment, such as moisture and temperature (See Chapter 6).

On the basis of the critical strains, Sakurai (1997a) proposed the hazard warning levels for assessing the stability of tunnels by measured displacements.

Chapter 5

Critical strains of rocks and soils

5.1 DEFINITION OF CRITICAL STRAIN OF GEOMATERIALS

Critical strain in compression is defined as:

$$\varepsilon_0 = \sigma_c/E \tag{5.1}$$

where ε_0 is the critical strain, σ_c is the uniaxial compressive strength, and E is Young's modulus (Sakurai, 1981).

The reciprocal number of the critical strain was defined as the "modulus ratio", as proposed by Deere (1968). Even though the critical strain has a clear physical meaning, as shown in Figure 5.1, the physical meaning of the modulus ratio is unclear.

For tensile strains, Stacey (1981) proposed a critical extension strain for massive brittle rocks to assess slabbing or spalling failure.

Various kinds of rock and soil specimens were tested in the laboratory to determine the critical strain, and the results are plotted in relation to the uniaxial compressive strength of intact rocks, as shown in Figure 5.2.

It is of interest that the critical strain is continuously distributed from soils to hard rocks, as seen in the Figure. This implies that it is not necessary to classify the geomaterials as soils or rocks. If we know their uniaxial strength, the lower and upper

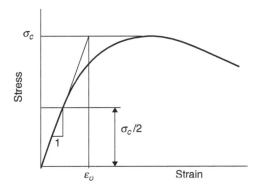

Figure 5.1 Definition of critical strain.

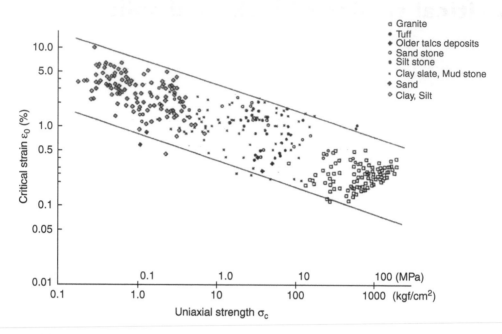

Figure 5.2 Relation between critical strain and uniaxial compressive strength of rocks and soils (Sakurai, 1981).

bounds of the critical strain can be easily evaluated, no matter whether the materials are soils or rocks.

5.2 SCALE EFFECT OF CRITICAL STRAINS

The critical strains, shown in Figure 5.2, are determined from small specimens of soils and intact rocks tested in the laboratory. In tunnel engineering practice, however, we need the critical strain of *in situ* rock masses, not of small-sized specimens. Thus, a question may arise how we can determine the critical strain of rock masses from the results of small specimens. In order to answer the question, we performed the *in situ* tests, such as plate bearing tests and direct shear tests, which were conducted at a tunnel construction site, where rocks consist of highly jointed granite. The results of both the plate bearing tests and the direct shear tests can provide Young's modulus E_R and the strength parameters, such as cohesion c and internal friction angle φ, assuming a linear Mohr-Coulomb's failure criterion. Once c and φ have been determined, the uniaxial compressive strength of the rock masses σ_{cR} can be calculated by the following equation:

$$\sigma_{cR} = \frac{c(1 - \sin \varphi)}{\cos \varphi} \qquad (5.2)$$

Table 5.1 Reduction coefficients for uniaxial compressive strength m and Young's modulus n (Sakurai, 1983).

Rock type & class	Rock core			Rock mass			Core vs mass		
	σ_c MPa	E_c GPa	ε_0 %	σ_{cR} MPa	E_m GPa	ε_{OR} %	m	n	m/n
granite CH	227.4	62.7	0.362	16.70	2.65	0.631	0.0734	0.0422	1.74
granite CM	211.5	59.8	0.534	11.60	1.96	0.592	0.0548	0.0328	1.67
granite CL	141.8	47.1	0.301	6.39	1.37	0.466	0.0451	0.0292	1.54
diorite CH	145.2	37.3	0.389	16.70	2.65	0.631	0.1150	0.0711	1.62
diorite CM	153.5	38.2	0.402	11.60	1.96	0.592	0.0755	0.0513	1.47
granite B	130.2	47.5	0.274	20.97	8.60	0.244	0.161	0.181	0.89
granite CH	117.6	44.3	0.265	14.31	2.98	0.480	0.198	0.067	1.81
granite CM	44.7	23.7	0.189	6.99	1.21	0.578	0.156	0.051	3.06
sandstone ---	137.2	28.4	0.483	6.27	0.381	1.642	0.046	0.0134	3.43
shale ---	78.4	21.6	0.364	3.82	0.411	0.930	0.049	0.0191	2.57
sandstone	137.8	54.1	0.255	7.89	1.96	0.403	0.057	0.0362	1.57
sandstone	22.9	10.1	0.227	2.49	1.08	0.231	0.109	0.107	1.01
shale	61.1	57.2	0.107	7.81	2.25	0.347	0.128	0.0393	3.26

The relationship between the critical strain of rock masses and intact rocks is given in Equation (5.3), with the result that the critical strain of rock masses ε_{0R} can be determined from the critical strain ε_0 of intact rock specimens.

$$\varepsilon_{0R} = \frac{\sigma_{cR}}{E_R} = \frac{m\sigma_c}{nE} = \left(\frac{m}{n}\right)\varepsilon_0 \qquad (5.3)$$

where σ_c: uniaxial compressive strength of intact rocks
  σ_{cR}: uniaxial compressive strength of rock masses
  E: Young's modulus of intact rocks
  E_R: Young's modulus of rock masses
  m: reduction coefficient for uniaxial compressive strength ($0 < m \leq 1.0$)
  n: reduction coefficient for Young's modulus ($0 < n \leq 1.0$)

In order to determine the reduction coefficients, intact rock specimens were drilled from rock masses at the place where both the plate bearing tests and the direct shear tests had been conducted. The intact rock specimens were tested in the laboratory to determine Young's modulus and the uniaxial compressive strength. Using the laboratory and *in situ* test results, we can determine the critical strains for both the intact rocks and the *in situ* rock masses obtained from the identical place, resulting in the reduction coefficients being determined, as shown in Table 5.1 (Sakurai, 1983).

It is obvious from this table that the ratio *m/n* is greater than 1.0, ranging approximately from 1.0 to 3.0. This means that the critical strain of jointed rock masses is approximately 1 to 3 times greater than that of intact rocks. This is the scale effect of critical strains, which indicates an entirely opposite trend to both the uniaxial compressive strength and Young's modulus. It is obvious that both the uniaxial compressive

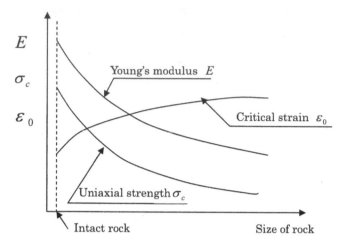

Figure 5.3 Scale effect on Young's modulus, uniaxial compressive strength and critical strain of rocks.

strength and Young's modulus of rock masses are always smaller than those of intact rocks, because of the fact that rock masses contain various joint systems in them, while it is interesting that the critical strain of rock masses is always larger than that of intact rock specimens, as schematically shown in Figure 5.3.

It is seen from the Figure that both the uniaxial compressive strength and Young's modulus of rocks decrease with an increase in size, because of a large specimen containing a large number of joints. This must be the reason why laboratory tests on intact rocks are unpopular in rock engineering practice, while in soil engineering practice, laboratory tests on small specimens are popularly used, because the mechanical characteristics of soils are more or less independent of the size of the specimens. However, if we take the ratio of the uniaxial compressive strength to Young's modulus, i.e. the critical strains, the effects of the joints more or less cancel each other out, resulting in critical strains slightly increasing with an increase in the size. This is a big advantage in rock engineering practice, because we can estimate the critical strain of jointed rock masses from laboratory tests on small specimens.

The critical strain of rock masses ε_{0R} is plotted in relation to the uniaxial compressive strength of rock masses σ_{cR} as shown in Figure 5.4. In this Figure, the critical strain of intact rocks is also plotted and connected with that of *in situ* rock masses, because both sets of data were obtained from the identical rock masses. The two dotted lines given in the Figure denote the upper and lower bounds of scattering data of critical strains shown in Figure 5.2. It is of interest to know from this Figure that the critical strain of rock masses is scattered in-between the upper and lower lines of the critical strain of the intact rock, and that the data for the intact rock seem to move parallel to the dotted lines towards those for the rock masses with a decrease in uniaxial compressive strength.

It is obvious from Figure 5.4 that the critical strain of rock masses is always greater than that of intact rocks. So, if we use the critical strain of intact rocks as an allowable

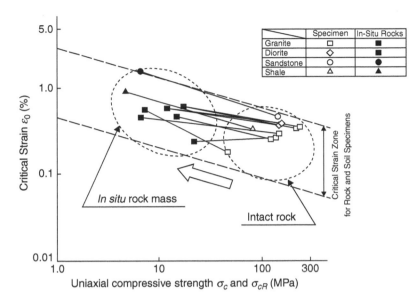

Figure 5.4 Relationship between the critical strain of intact rocks and *in situ* rock masses.

value (hazard warning levels), a certain amount of margin of safety can always be guaranteed in assessing the stability of structures. In conclusion, the critical strain of intact rocks can be directly used for assessing the stability of structures, while both the uniaxial strength and Young's modulus of intact rocks cannot be used in engineering practice, unless the scale effect is being properly considered.

5.3 SIMPLE APPROACH FOR ASSESSING TUNNEL STABILITY

An approach for assessing the stability of tunnels should be simple enough to be easily applied for tunnelling practice at construction sites. For this purpose, it is assumed that a tunnel is approximately circular in shape and has been excavated in ground consisting of continuous and homogeneous geomaterials with the hydrostatic state of initial stresses. According to two-dimensional continuum solid mechanics, circumferential strain ε_θ around a circular tunnel is expressed as:

$$\varepsilon_\theta = \frac{u}{r} \tag{5.4}$$

where u: radial displacement of the tunnel
    r: radial coordinate

In monitoring the stability of tunnels, the crown settlements and the convergences are commonly measured during the excavation. Thus, considering Equation (5.4), we can determine the circumferential strain at the inner surface of the tunnels either from the crown settlements or the convergences, as follows:

$$\varepsilon_{\theta, r=a} = \frac{\delta}{a} \tag{5.5}$$

where δ: crown settlements or convergences/2
  a: tunnel radius

Now, let us show a case study. A double-lane freeway tunnel was excavated in heavily jointed granite and welded tuff with many fractures and shear zones. The crown settlements and the convergences were measured during the excavation. The strain occurring around the tunnel was calculated by Equation (5.5). The results obtained by the crown settlements were plotted in relation to the uniaxial compressive strength of intact rocks, as shown in Figure 5.5.

Strains were also measured by borehole extensometers installed from the tunnel surface, and the results were plotted in relation to the uniaxial compressive strength of intact rocks, as shown in Figure 5.6.

In these two Figures, the data indicated by the dark circles were obtained at the tunnel sections where serious trouble occurred during the excavation, while the white circles were obtained at the tunnel sections where the excavations were completed without any trouble. The associated numbers indicate the sort of trouble, as given in

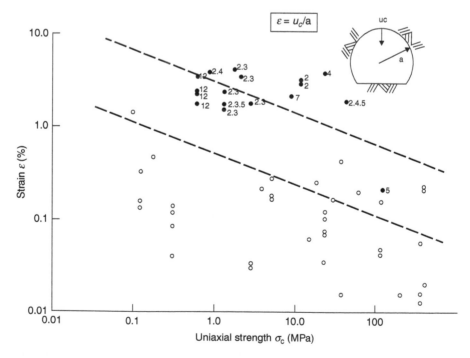

Figure 5.5 Relationship between measured strain (obtained from crown settlements) and the upper and lower bounds of critical strain (dotted lines) (Sakurai, 1997a).

Table 5.2. The two dotted lines indicate the upper and lower bounds of the critical strain, respectively, as shown in Figure 5.2.

It is obvious from Figures 5.5 and 5.6 that if the measured strains remain below the centre of the two dotted lines, the tunnels were excavated without any problems, while if the measured strains are above the middle line, some troubles occurred during the excavations. Furthermore, if the measured strain surpasses the upper line, a lot of serious troubles surely occurred.

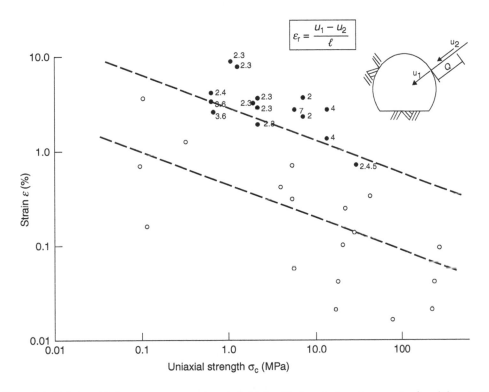

Figure 5.6 Relationship between measured strain (obtained by borehole extensometers) and the upper and lower bounds of critical strain (dotted lines) (Sakurai, 1997a).

Table 5.2 Problems encountered during tunnelling (number indicated in the table corresponding to the numbers indicated in the previous Figures).

No.	Remarks
1	Difficulty in maintaining tunnel face
2	Failure and/or cracking in shotcrete
3	Buckling of steel ribs
4	Breakage of rock bolts
5	Fall-in of roof
6	Swelling at invert
7	Miscellaneous (unidentified)

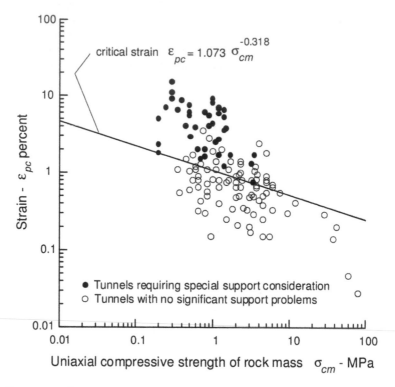

Figure 5.7 Comparison between measured strain (obtained from crown settlements) and the centre line of the scattered values of critical strains (Hoek, 1998).

Another example for assessing the stability of tunnels by the critical strain is shown in Figure 5.7, where the strains were determined by measured crown settlements. The line shown in the Figure is the centre line (average value) of the scattering data of critical strains shown in Figure 5.2. It is obvious from this Figure that the data above the centre line of the critical strains show that the tunnels required special support countermeasures, while under the centre line no significant support problems occurred. This is the same conclusion as the tunnels shown in Figures 5.5 and 5.6 (Hoek, 1998).

(Information shown in Figure 5.7 was supplied by Dr J. C. Chern of Sinotech Engineering Consultants Inc., Taipei. Taiwan)

5.4 HAZARD WARNING LEVEL FOR ASSESSING CROWN SETTLEMENTS AND CONVERGENCE

Considering the results shown in Figures 5.5 and 5.6, Sakurai (1997a) proposed hazard warning levels for assessing the stability of tunnels. The hazard warning levels are classified into three stages depending on the level of tunnel stabilities. In addition, the

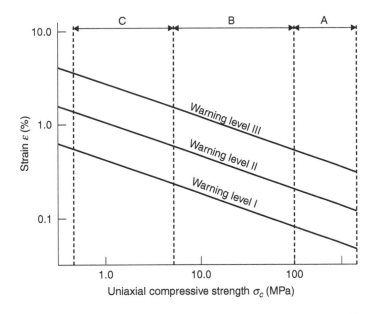

Figure 5.8 Hazard warning levels in terms of strain in tunnels (Sakurai, 1997a).

uniaxial compressive strength of geomaterials surrounding tunnels are also classified into three groups (A, B, and C), as shown in Figure 5.8.

In tunnel construction sites, it is extremely important to assess the stability of tunnels immediately after taking the data of measured displacements. For this purpose, the hazard warning levels expressed in terms of strains shown in Figure 5.8 are converted into those expressed in terms of displacements, i.e. crown settlements and convergences, considering Equation (5.5) as shown in the following equation.

$$d = \varepsilon_0 a \tag{5.6}$$

where d: hazard warning level for radial displacements at the inner surface of tunnels
  ε_0: critical strain of geomaterials
  a: tunnel radius

As an example, the hazard warning levels for the crown settlement of tunnels with a 5 m radius are derived from Equation (5.6), and shown in Table 5.3. By the use of this Figure, the tunnel stability can be assessed in terms of crown settlements, immediately after taking the data.

In tunnel constructions, the crown settlements are usually measured for monitoring the stability of tunnels during excavations. However, the measured settlements are only part of total settlements, because a certain amount of displacements has already taken place when the measurements start, so that the measured values of the settlements cannot be compared with the hazard warning levels shown in Table 5.3, unless the displacements which are not measured are added to the measured displacements. For

Table 5.3 Hazard warning levels for the crown settlements of tunnels with 5 m radius.

	A	B	C
I	0.3~0.5	0.5~1	1~3
II	1~1.5	1.5~4	4~9
III	3~4	4~11	11~27

(Unit: cm)
(Radius of tunnel: 5.00 m)

this purpose, numerical simulations must be useful for predicting the percentage of the settlements that has already taken place when the settlement measurements start, and the total crown settlements can be determined as the sum of the measured settlements and the numerically simulated settlements.

It is noted that the critical strain of rock masses is always larger than that of intact rocks, as shown in Figure 5.4. Therefore, if we use the critical strain of intact rocks for assessing the stability of tunnels, it always allows a certain amount of margin for the safety of tunnels. This amount of margin of the displacements tends to compensate with the displacements which have already taken place when the measurements start. Since all the measurement data shown in Figures 5.5 and 5.6 represent the strains measured after measurements started, they are not the total strains but part of the total strains, whereas the measured strains are consistent with the criterion of critical strains as shown in Figures 5.5 and 5.6. This may be due to the fact that the uniaxial compressive strength of geomaterials shown in the Figures was determined by laboratory tests on intact rocks, so that the quantity of measured displacements corresponds to the part of the total displacements.

Now, there are some cautions for tunnel engineers. When the tunnels are excavated through the boundary of different geological formations from soft to hard rocks, tunnel engineers should be careful to assess the stability of tunnels in terms of displacements such as crown settlements and convergences, due to the fact that the displacements become smaller than that of soft ground. As a result, the engineers may feel that the tunnels tend to be stable, whereas the hazard warning levels also become smaller as the uniaxial compressive strength increases, as shown in Figure 5.8. In other words, it is obvious that hard rocks fail in smaller strains than soft rocks. Therefore, when the tunnels are excavated through the boundary from soft rocks to hard rocks, there is a possibility that the tunnel crown may cave in, even though the displacements become small.

5.5 UNIAXIAL COMPRESSIVE STRENGTH AND YOUNG'S MODULUS OF ROCK MASSES

The critical strain of rocks and soils is related to their uniaxial compressive strength, as shown in Figure 5.2, whereas, according to the definition of the critical strain given in Equation (5.1), the critical strain can also be expressed in relation to Young's modulus, as shown in Figure 5.9 (Sakurai, 1990).

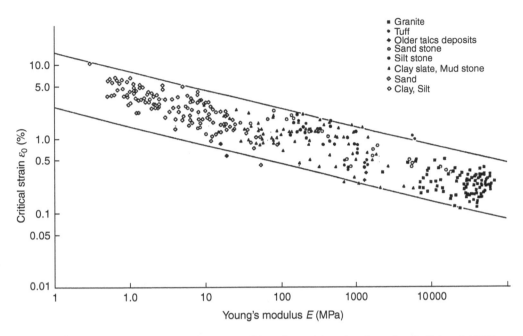

Figure 5.9 Relation between critical strain and Young's modulus of rocks and soils (Sakurai, 1990).

Even though all the data shown in this Figure were obtained by the laboratory experiments on small specimens, this Figure is also valid for *in situ* rock masses, as shown in Figure 5.4.

In the design analysis of tunnels, the mechanical parameters such as Young's modulus and strength parameters (cohesion and internal friction angle) of rock masses are important input data, which can be determined by plate bearing tests and direct shear tests. However, they are costly so that they are not commonly performed in engineering practice. To avoid such a high cost of *in situ* tests, we often use the borehole loading tests, which are most popularly performed for determining the deformability (Young's modulus) of rock masses.

On the other hand, there are no reasonable *in situ* tests for determining the strength parameters, such as cohesion and internal friction angle of rock masses. To overcome the difficulties in determining the strength parameters of rock masses, Figure 5.9 is used, in such a way that if Young's modulus of rock masses is determined by borehole loading tests, then the critical strain of rock masses ε_{cR} can be approximately evaluated by Figure 5.9, resulting in uniaxial compressive strength of rock masses σ_{cR} being determined by the definition of critical strains, as shown in Equation (5.7):

$$\sigma_{cR} = \varepsilon_{0R} E_R \tag{5.7}$$

where ε_{0R}: critical strain of rock masses
$\quad\quad E_R$: Young's modulus of rock masses

If we assume the Mohr-Coulomb yielding criterion, uniaxial strength σ_{cR}, cohesion c and internal friction angle φ are related to each other, as follows:

$$c = \frac{\sigma_{cR}(1 - \sin\varphi)}{2\cos\varphi} \tag{5.8}$$

Equation (5.7) together with Equation (5.8) can be used for confirming an inconsistency among the input data, such as Young's modulus and strength parameters (cohesion and internal friction angle), used in the design analysis of structures. This must be very useful for improving reliability of the input data for the design analyses, because no information on field measurement data is available at the design stage. Nevertheless, once excavations start, displacement measurements can be performed, followed by back analyses being carried out for determining Young's modulus of rock masses by using measured displacements (Sakurai & Takeuchi, 1983), whereas it is hard to back-calculate the strength parameters of rock masses, even though the back analysis technique is used. To solve this problem, Equations (5.7) and (5.8) are used to determine the strength parameters from the back-calculated Young's modulus.

It is noted that in tunnelling projects, Young's modulus and the strength parameters, such as cohesion and internal friction angle of rock masses, are often determined by different engineer groups, resulting in that there is some possibility for the results of the mechanical parameters being inconsistent with each other. To overcome this problem, design engineers should always confirm whether Young's modulus and the strength parameters are consistent with each other. For this purpose, Equations (5.7) and (5.8) are useful for avoiding the inconsistency among the input data used in the design analyses.

Chapter 6

Environmental effects on critical strain of rocks

6.1 CRITICAL STRAIN IN TRIAXIAL CONDITION

The critical strain is originally defined under a uniaxial condition. In engineering practice, however, geomaterials are usually loaded under triaxial condition. Thus, the critical strain in triaxial condition is defined as follows (see Figure 6.1) (Sakurai et al., 1994b):

$$\varepsilon_0 = \frac{(\sigma_1 - \sigma_3)_f}{E} \tag{6.1}$$

where $(\sigma_1 - \sigma_3)_f$ denotes maximum stress deviator at failure, and E is Young's modulus.

6.2 EFFECTS OF CONFINING PRESSURE

Porous tuff was tested under triaxial conditions in the laboratory. The size of a cylindrical specimen was ϕ50 mm $\times$ H100 mm. The results of tests under both fully-saturated and drained conditions are shown in Figures 6.2, 6.3 and 6.4, where maximum stress deviator, Young's modulus and critical strain are plotted with respect to confining pressure, respectively. It is clearly understood from these Figures that the maximum stress deviator and Young's modulus are strongly influenced by confining pressure and they increase with an increase in confining pressure, while the critical strain is fairly independent from confining pressure, and indicates an almost constant value. It is worth mentioning that the critical strain under the triaxial condition is almost identical to that of the uniaxial condition (confining pressure is zero) (Sakurai et al., 1994b).

It is noted that confining pressures slightly influence the critical strain when the confining pressure increases, as shown in Figure 6.5. Looking at the Figure, we can see that the critical strain is influenced by confining pressure becoming greater than 5 MPa, whereas the effect of confining pressure may be neglected in tunnel engineering practice, because the minimum principal stress in rocks near the inner surface of tunnels must be not large due to the excavations. Therefore, the value of critical strain determined under uniaxial compressive tests can be used for assessing the stability of the tunnels.

In fact, Kohmura (2012) concluded that the stability of tunnels can be assessed by using the critical strain determined by the uniaxial compressive tests, resulting in the stability assessment being safe side.

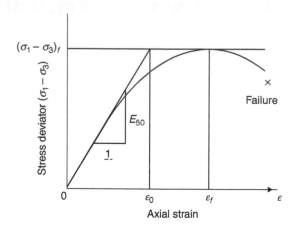

Figure 6.1   Definition of critical strain in triaxial condition.

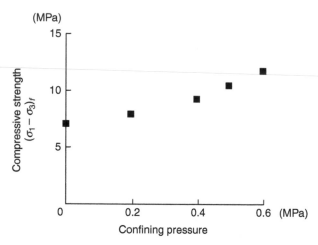

Figure 6.2   Effect of confining pressure on compressive strength (porous tuff).

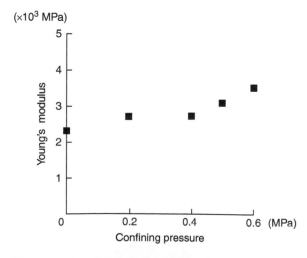

Figure 6.3   Effect of confining pressure on Young's modulus (porous tuff).

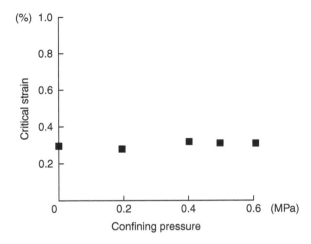

Figure 6.4 Effect of confining pressure on critical strain (porous tuff).

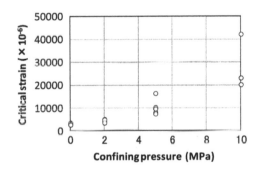

Figure 6.5 Effect of confining pressure on critical strain of tuff under high confining pressure (Kohmura, 2012).

6.3 EFFECTS OF MOISTURE CONTENT

Cylindrical specimens of porous tuff were also used. The tests were carried out under a uniaxial compressive stress state by changing moisture content ranging from 0% to 100% saturation with 20% interval. Three tests were repeatedly conducted for each moisture content (Sakurai et al., 1994b).

The relationship between uniaxial compressive strength and moisture content is shown in Figure 6.6. Young's modulus is shown in Figure 6.7 for different moisture contents. It can be observed from these two Figures that both the uniaxial compressive strength and Young's modulus decrease with an increase in moisture content. On the other hand, however, it is understood from Figure 6.8 that the critical strain seems to be an approximately constant value with the exception of the under low moisture content condition. It may be of interest to note that the critical strain tends to increase

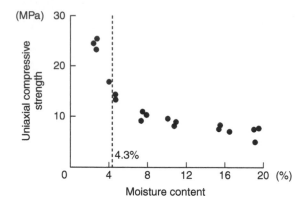

Figure 6.6   Effect of moisture content on uniaxial compressive strength.

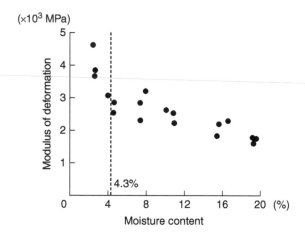

Figure 6.7   Effect of moisture content on modulus of deformation.

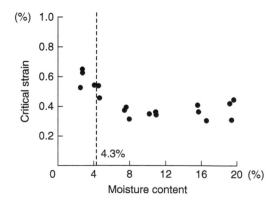

Figure 6.8   Effect of moisture content on critical strain.

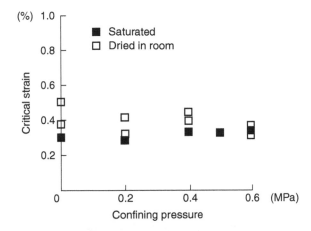

Figure 6.9 Effect of confining pressure on critical strain.

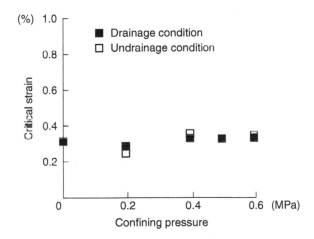

Figure 6.10 Effect of confining pressure on critical strain (triaxial compressive state).

rapidly as moisture content becomes less than the natural moisture of the laboratory (the natural moisture was 4.3% during tests).

Figure 6.9 shows a comparison between critical strains of both fully-saturated and dried specimens determined under different confining pressures. It can be seen here that both results are almost identical to each other. This indicates that the effect of confining pressure is negligibly small even for saturated rocks.

Triaxial tests were also carried out under both drainage and non-drainage conditions. The results are shown in Figure 6.10. It can be deduced from this Figure that pore water condition (drainage or non-drainage) and confining pressure do not cause any influence on critical strain. Therefore, it may be concluded that critical strain is almost independent from the change of both pore water condition and confining pressure.

(a)   Saturated

(b)   Dried in room

(c)   Absolutely dried

Photo 6.1 Microstructure of specimen (porous tuff)(×2,000) (Sakurai et al., 1994b).

In order to verify the reason why the critical strain increases when moisture content is less than natural moisture, microstructures of the rocks were investigated by using a scanning electron microscope. The microscopic structures of rocks are shown in photos 6.1(a), (b) and (c). The photos 6.1(a) and (b) are taken for the rocks under

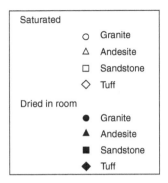

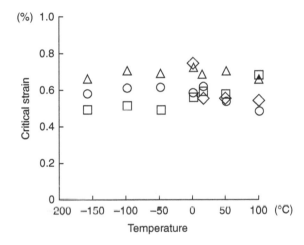

Figure 6.11 Effect of temperature on critical strain (saturated specimens) (Inada & Kokudo, 1992).

fully-saturated and natural moisture conditions, respectively. It is seen from these photos that there seems to be no difference between the two. On the other hand, photo 6.1(c) is for a rock in an absolutely dried condition. The photo indicates clearly the change of microstructure of the rocks due to an absolute dryness.

The comparison of these three photos may conclude that the physical constitution of the rocks is no more the same as the original one, when moisture content of the rocks becomes less than the natural moisture in a room. In other words, the rocks become different types, so that critical strain is no more constant.

6.4 EFFECTS OF TEMPERATURE

The effects of temperature on critical strain were investigated by using data already published in literatures. One of the literatures shows the results of the uniaxial compression tests on granite, andesite, sandstone and tuff under different temperatures

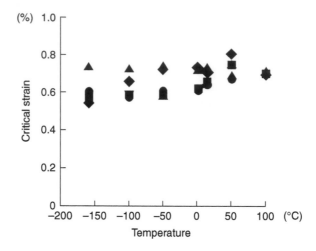

Figure 6.12   Effect of temperatures on critical strain (dried specimens) (Inada & Kokudo, 1992).

(Inada & Kokudo, 1992), whose test results are introduced here. The critical strains are determined and plotted with respect to temperature, as shown in Figure 6.11 (fully-saturated) and Figure 6.12 (dried specimens). It may be of interest to note that critical strain is also independent from temperature as well as moisture content.

Chapter 7

General approach for assessing tunnel stability

7.1 CRITICAL SHEAR STRAIN OF GEOMATERIALS

As described in Section 5.3, the stability of tunnels can be assessed during the excavation by comparing the measured crown settlements and/or convergences with the hazard warning levels derived from the critical strain of the geomaterials. This assessment approach is so simple and practical that the practising engineers can get some idea of the overall stability of tunnels during the excavations, although it is not precise enough for all cases, because Equation (5.5) is only valid for tunnels with a circular cross section excavated in the ground under a hydrostatic initial state of stress.

In order to assess the stability of tunnels in a more precise manner, Sakurai et al. (1993a) proposed the critical shear strain of geomaterials. The tunnel stability can then be assessed by comparing the maximum shear strain occurring in the ground around the tunnels with the critical shear strain, in the same fashion as for assessing both crown settlements and convergences in tunnels on the basis of the critical strain under a compression state, as described in Figure 5.8.

The critical shear strain is defined as the ratio of the maximum shear strength to the shear modulus, as shown in Equation (7.1), in a similar fashion to the critical strain given in Equation (5.1).

$$\gamma_0 = \tau_c/G \tag{7.1}$$

where γ_0: critical shear strain
τ_c: maximum shear strength
G: shear modulus

The critical shear strain of soils can be determined by torsion tests using a hollow cylindrical specimen in the laboratory. For rocks, however, the torsion tests can rarely be adopted because of the difficulty in testing. To avoid this difficulty, the critical shear strain is converted from the critical strain, using Equation (7.2), which is derived under the assumption of the materials remaining in an elastic state until the axial strain reaches the critical strain.

$$\gamma_0 = (1 + \nu)\varepsilon_0 \tag{7.2}$$

where γ_0: critical shear strain
ν: Poisson's ratio
ε_0: critical strain in compression

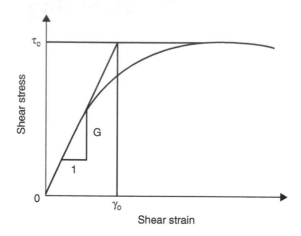

Figure 7.1 Definition of critical shear strain.

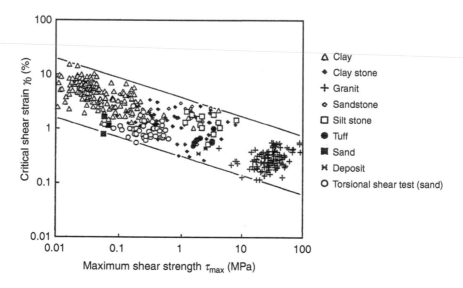

Figure 7.2 Relationship between critical shear strain and maximum shear strength (Sakurai et al., 1993a).

The critical shear strain of rocks and soils can then be converted from the critical strain shown in Figure 5.2 by using the conversion Equation (7.2), and the results are plotted in relation to the maximum shear strength, as shown in Figure 7.2.

In order to verify the adequacy of Equation (7.2), torsion tests on cohesive soils using a hollow cylindrical specimen were carried out to determine the critical shear strain, and the results are plotted in Figure 7.2. It is obvious from the figure that all the data obtained in the two different ways, i.e. the conversion from the critical strain using Equation (7.2), and the torsion tests performed on cohesive soils using a

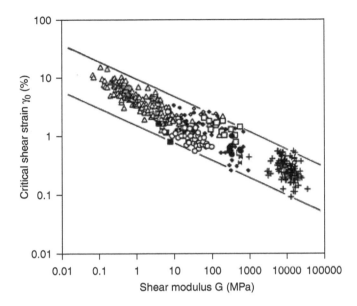

Figure 7.3 Critical shear strain of various rocks and soils in relation to the shear modulus of the geomaterials (Sakurai et al., 1993a).

hollow cylindrical specimen, fall into the scattering data between the upper and lower bounds. This proves that Equation (7.2) is suitable for converting the critical strain to the critical shear strain with high reliability.

In engineering practice, however, it is difficult to determine the strength parameters, such as the maximum shear strength of *in situ* rocks, from measured displacements. Instead of the maximum shear strength therefore the deformability, such as shear modulus, is much more easily determined from measured displacements. Thus, it is preferable to plot the critical shear strains in terms of the shear modulus, as shown in Figure 7.3. It is noted that the shear modulus can be easily back-calculated from measured displacements, as described in Section 21.2.

7.2 HAZARD WARNING LEVELS IN TERMS OF MAXIMUM SHEAR STRAIN

As already described, the stability of tunnels can be assessed by comparing measured crown settlements and convergences with the hazard warning levels derived from the critical strain of rocks and soils (see Figure 5.8). This stability assessment approach can easily be applied for monitoring the overall stability of tunnels during constructions, but it is not precise enough to assess the tunnel stability; for example, deformational behaviour around a shallow tunnel is often characterised by the formation of shear bands developing from the tunnel shoulder reaching, sometimes, to the ground surface. As an example, Figure 7.4 shows the maximum shear strain distribution derived from

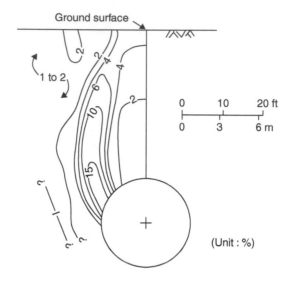

Figure 7.4 Maximum shear strain distribution (Hansmire & Cording, 1985).

the displacement measurement data taken from a subway tunnel in Washington DC (Hansmire & Cording, 1985).

In order to assess the tunnel stability in a more precise manner, the critical shear strain is used in such a way that the maximum shear strains occurring around tunnels are determined from measured displacements by back analyses, and compared with the critical shear strains of rocks. For this purpose, the hazard warning levels are proposed in terms of the maximum shear strain as shown in Figure 7.5, in the same manner as for tunnel crown settlements and convergences (see Figure 5.8). In the figure, the horizontal axis is the shear modulus, and the vertical axis is the maximum shear strain. Both can be back-calculated from the displacements measured during the excavations.

Let us now explain how to use the figure for assessing the stability of tunnels. As an example, we will assume the maximum shear strain occurring around the tunnel being back-calculated to be 1.5%, and the shear modulus of the surrounding rock mass is back-calculated as 10 MPa. As seen in Figure 7.5, the point of the intersection of both the maximum shear strain of 1.5%, and the shear modulus of 10 MPa, lies near the centreline (Warning Level II), but a little lower than that, as shown in the figure. This indicates that the stability of the tunnel gets close to Warning Level II, but the tunnel is still stable at present.

It is noted that, if the maximum shear strain occurring around a tunnel is smaller than the line of Warning Level II, no plastic zone has yet appeared, resulting in the tunnel being stable. The centreline may correspond to the elastic limit of the geomaterials, due to the fact that no additional support measures are needed, provided that the measured strains are still under the centreline, as shown in Figures 5.5, 5.6 and 5.7. If the maximum shear strain becomes greater than the line of Warning Level II, some additional support measures should be installed.

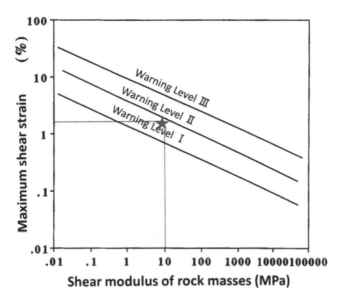

Figure 7.5 Hazard warning levels for assessing the stability of difficult tunnels.

This approach for assessing the stability of tunnels has been successfully used during the excavation of the Sirkeci station in the Marmaray Project in Istanbul, Turkey (Iwano et al., 2010; Otsuka et al., 2011).

7.3 HOW TO DETERMINE THE MAXIMUM SHEAR STRAIN DISTRIBUTION AROUND A TUNNEL

If the amount of displacement measurement data is sufficiently large, the strains can be determined directly from the measured displacements by using the kinematic relationship between strains and displacements without assuming any mechanical model of rock masses, as follows:

$$\{\varepsilon\} = [B]\{u\} \tag{7.3}$$

where $\{\varepsilon\}$ is strains, $\{u\}$ is measured displacements, and $[B]$ is a matrix which is only a function of the location of measurement points.

Sakurai (1981, 1982) proposed a method for determining strain distributions around tunnels by the kinematic relationship alone. The detail of the method is described as follows:

Let us consider the displacements around a tunnel measured by multi-rod borehole extensometers as well as inclinometers at several points around the tunnel, as shown in Figure 7.6. If the ground has no major joints, the displacement distributions can be expressed in a continuous function, and the strain distributions around the tunnel can then be derived by taking derivatives of the displacement function considering the

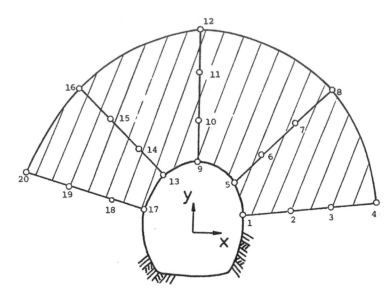

Figure 7.6 Measurement points around a tunnel.

relevant kinematic relationships in continuum mechanics. The displacements in the region surrounded by the measurement points (shaded regions shown in Figure 7.6) can be interpolated in terms of the displacements measured at the measuring points. The region surrounded by the measurement points is called an "element".

The shape of an element can be expressed in terms of interpolation functions and the coordinates of the measuring points, i.e.,

$$x = \sum_{i=1}^{N} P_i(\xi, \eta) x_i$$

$$y = \sum_{i=1}^{N} P_i(\xi, \eta) y_i \tag{7.4}$$

where $P_i(\xi, \eta)$ is the interpolation function in terms of the curvilinear coordinates ξ and η. x_i and y_i are, respectively, the x and y coordinates of measurement points i, and N is the total number of measurement points in an element.

If we assume the isoparametric element proposed by Zienkiewicz (1971), the displacements in an element are expressed in terms of the same interpolation functions as the shape function. Hence, the displacements are given by:

$$u = \sum_{i=1}^{N} P_i(\xi, \eta) u_i$$

$$v = \sum_{i=1}^{N} P_i(\xi, \eta) v_i \tag{7.5}$$

where u_i and v_i are the measured displacements in the x and y directions at the measuring point i.

In the two-dimensional case, the relationship between displacements and strains is given as follows:

$$\varepsilon_x = \frac{\partial u}{\partial x}, \quad \varepsilon_y = \frac{\partial v}{\partial y}, \quad \gamma_{xy} = \frac{\partial u}{\partial y} + \frac{\partial v}{\partial x} \tag{7.6}$$

Substituting Equation (7.5) into Equation (7.6) gives the strain distribution in an element in terms of the displacements measured at the measuring points, i.e.,

$$\{\varepsilon\} = [B]\{u\} \tag{7.7}$$

where
$\{\varepsilon\} = <\varepsilon_x \quad \varepsilon_y \quad \gamma_{xy}>$: strain vector in the area surrounded by measuring points
$\{u\} = <u_1 \quad v_1 \quad u_2 \quad v_2.....u_N \quad v_N>$: measured displacements at measuring points
$[B] = [T_1(\xi, \eta)..T_2(\xi, \eta)..T_N(\xi, \eta)]$: matrix of the relationship between strain and displacement

where

$$[T_i(\xi, \eta)] = \begin{bmatrix} \dfrac{\partial P_i}{\partial x} & 0 \\ 0 & \dfrac{\partial P_i}{\partial y} \\ \dfrac{\partial P_i}{\partial y} & \dfrac{\partial P_i}{\partial x} \end{bmatrix} \tag{7.8}$$

Since the interpolation function $P_i(\xi, \eta)$ is given by a curvilinear coordinate system, the coordinate transformation is necessary to calculate the derivatives of Equation (7.8). The principal strains can be obtained as follows:

$$\begin{Bmatrix} \varepsilon_1 \\ \varepsilon_2 \end{Bmatrix} = \frac{\varepsilon_x - \varepsilon_y}{2} \pm \sqrt{\left(\frac{\varepsilon_x - \varepsilon_y}{2}\right)^2 + \frac{1}{4}\gamma_{xy}^2} \tag{7.9}$$

where ε_1: maximum principal strain
$\quad \varepsilon_2$: minimum principal strain
The maximum shear strain γ_{max} can then be calculated as follows:

$$\gamma_{max} = |\varepsilon_1 - \varepsilon_3| \tag{7.10}$$

This equation shows that the maximum shear strain distributions around a tunnel can be determined from measured displacements only by using a kinematic relationship. The above-mentioned two-dimensional approach can be easily extended to a three-dimensional case (Sakurai, 1982).

The stability of tunnels can be assessed by comparing the occurring maximum shear strain given in Equation (7.10) with the hazard warning levels in terms of the

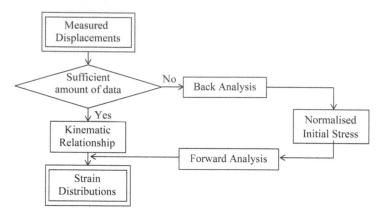

Figure 7.7 Flow chart for two different approaches determining strain distributions around tunnels from measured displacements.

critical shear strain of rocks and soils, shown in Figure 7.5. This approach is advantageous, because no information is needed with respect to both the initial stress of the ground and the mechanical parameters of the geomaterials. In engineering practice, however, the number of extensometers installed around the tunnels is in general not large enough to determine the strain distributions over a large extent of the ground around the tunnels. Therefore, the above-mentioned method for determining the strain distributions can be hardly applicable to engineering practice.

To overcome this difficulty, a back analysis procedure can be used for determining the maximum shear strain distributions from the measured displacements. According to the back analysis procedure, the strain distribution around the tunnels can be determined only by a limited amount of measurement data, so that the back analysis procedure has a great advantage for engineering practice.

It is obvious from the above discussion that there are two approaches available for determining the strain distributions around tunnels from measured displacements. One is a direct method in which the strains are obtained directly by using the kinematic relationship between strain and displacement. The other is an indirect method in which the "normalised initial stresses" of the ground (see Chapter 8) are first determined by the back analyses of measured displacements, and are then used as input data for a forward analysis to calculate the strain distributions around tunnels.

Figure 7.7 shows the two different approaches for determining the maximum shear strain distributions around the tunnels from measured displacements, either by the direct method using the kinematic relationship or the indirect method based on back analyses.

Chapter 8

Back analyses used in tunnel engineering practice

8.1 INTRODUCTION

In the design of tunnels, both mechanical parameters and initial stresses of rock masses are important input data. However, it is not an easy task to determine them rationally by laboratory and *in situ* tests, because of various uncertainties due to the complex geological and geomechanical characteristics of rock masses involved. To overcome these difficulties, observational methods are usually adopted during construction, in such a way that the input data used in the design analyses are verified by the back analysis of measured displacements, and modified if necessary. For this purpose, various types of back analysis procedures have been proposed, but most of them are to determine the mechanical parameters alone, while the initial stresses are assumed to be the same as those used in the design analyses. Nevertheless, the back analysis procedures in the observational methods should be capable of back-calculating both mechanical parameters and initial stresses of the ground.

It is noted that, in the observational methods, one of the important purposes of back analyses must be to assess the stability of rock structures during construction. For this purpose, Sakurai & Takeuchi (1983) proposed a back analysis procedure for assessing the stability of tunnels, where "normalised initial stress" was defined as the ratio of the initial stresses of rock masses to Young's modulus of geomaterials surrounding the tunnels. According to the proposed method, the normalised initial stresses are first determined by back analyses of measured displacements. Following this, the maximum shear strain distribution around tunnels can be calculated by a forward analysis using the back-calculated normalised initial stress as input data. As a result, the stability of the tunnels can be assessed by comparing the back-calculated maximum shear strains with the hazard warning levels of the geomaterials (see Figure 7.5).

The advantage of this back analysis method is that the stability of tunnels can be assessed by only using measured displacements during construction, without considering both the initial stress and Young's modulus of rock masses. If, however, each component of initial stress is needed, then it can be determined from the normalised initial stress by assuming the vertical component of initial stresses to be the same as the overburden pressure. Once the component of the initial stresses is determined, then Young's modulus is also calculated from the normalised initial stresses. Furthermore, in the proposed back analysis procedure, only three measurement data points are sufficient to determine the maximum shear strain distribution around tunnels in a deterministic way. If more than three data points are available, an optimisation technique, such as a least squares method, can be used to increase the reliability of the back analysis results.

Another important aspect of back analyses for assessing the stability of tunnels during the excavations is to achieve the prompt feedback of back analysis results to the way of excavation procedures, and to verify the adequacy of support measures without any delay. To satisfy this requirement, an inverse back analysis approach is preferable because it does not require an iteration process in calculation. In this respect, the proposed back analysis procedure formulated in an inverse approach has superiority for application to engineering practice. However, the mathematical formulation of the proposed back analysis procedure is based on a stiffness matrix method, which requires a large stiffness matrix of whole structure systems together with the inversion of the matrix, resulting in the computation time drastically increasing.

To overcome this problem, Sakurai & Shinji (1984) modified the back analysis procedure by using flexibility matrix methods, resulting in the computation time being significantly reduced because there is no need to carry out the inversion of the large stiffness matrix. The flexibility matrix methods are formulated by the same fundamental principles as those of the stiffness matrix methods, the only difference being that the flexibility matrix is used in the mathematical formulations instead of the stiffness matrix. As a result, the back analysis procedures can provide prompt feedback of the results of the back analyses to verify the adequacy of the original design of support measures during the excavations without any delay.

It is noted that the shortcoming of the proposed back analysis procedures is that the mechanical model of materials is assumed to be a linear elastic model, so that non-elastic behaviours such as plastic deformations, loosening due to the gravitational force, or crack openings due to blasting cannot be taken into account. As already mentioned, in the back analyses in rock engineering practice, the mechanical model should not be assumed, but should be determined by back analyses (Sakurai, 1997a). In this regard, it is necessary to improve the proposed back analysis method considering the non-elastic behaviour of geomaterials. For this purpose, both the normalised initial stress and non-elastic strains must be considered together in back analyses, resulting in back-calculated strain distributions including both elastic strains and non-elastic strains. The back analysis procedure considering non-elastic strains is described in Chapter 9.

8.2   MATHEMATICAL FORMULATION OF THE PROPOSED BACK ANALYSES

8.2.1   Introduction

In tunnel engineering practice, the main purpose of back analyses is not only to back-calculate the mechanical parameters of the geomaterials surrounding tunnels, but also to assess the stability of tunnels, in such a way that the maximum shear strain distributions around the tunnels are back-calculated from measured displacements, and compared with the hazard warning levels expressed in terms of the critical shear strain of the geomaterials, as shown in Figure 7.5. In this section, the mathematical formulation of the back analysis procedure based on the flexibility matrix methods proposed by Sakurai & Shinji (1984) is described, in which the normalised initial stress is used, so that the initial stress of the ground around the tunnels is not needed.

8.2.2 Assumption of mechanical model

In the mathematical formulation, the following assumptions are made:

1   The mechanical characteristics of soils and rocks are expressed in homogeneous isotropic linear elastic models, so that the material constants are Young's modulus and Poisson's ratio. However, since Poisson's ratio does not have much influence on the results of back analyses, an adequate value can be assumed.
2   Young's modulus of the lining materials (usually concrete) is assumed to be known.
3   The initial stress of rock masses is uniformly distributed throughout the ground being excavated.

The above-mentioned assumptions were used not only for the stiffness matrix method, but also for the flexibility matrix method.

8.2.3 Mathematical formulation

In finite element analyses for excavation problems of tunnels, the initial stresses of the ground are taken into account by applying equivalent nodal force $\{P_0\}$ at the excavation surface, which corresponds to the initial state of stress $\{\sigma_0\}$ in the ground being excavated. This force is determined by:

$$\{P_0\} = \int_V [B]^T \{\sigma_0\} dv + \int_V [N]^T \{p\} dv \tag{8.1}$$

where $\{p\}$ is the vector of the body force components due to the gravitational force
$[N]$ and $[B]$ are the matrices of element shape functions and their derivatives, respectively
v is excavation volume

It is noted that in the back analyses the mechanical model of materials is assumed as a linear elastic material, and the effect of the gravitational force in Equation (8.1) is disregarded because the tunnels are assumed to be excavated at a great depth. This assumption makes it possible to adopt the inverse approach, so that an iteration process can be avoided. However, in the case of tunnels at a shallow depth, the term of the gravitational force cannot be disregarded; hence the direct back analysis approach is used.

In this section, the two-dimensional formulation is presented, resulting in the following initial stress components:

$$\{\sigma_0\} = \{\sigma_{x0}, \quad \sigma_{y0}, \quad \tau_{xy0}\}^T \tag{8.2}$$

The relation between the nodal forces $\{P\}$ and the nodal displacements $\{u\}$ is expressed by the well-known relationship,

$$[K]\{u\} = \{P\} \tag{8.3}$$

where $[K]$ denotes the stiffness matrix of the assembled finite element system. For lined tunnels with Young's modulus E_R and E_L of rocks and linings respectively, the stiffness matrix is expressed as

$$[K] = E_R[K^*] \tag{8.4}$$

where $[K^*] = [K_R] + R[K_L]$

$$R = E_L/E_R$$

$[K_R]$ represents the stiffness matrix for the ground for $E_R = 1$, and $[K_L]$ that for the lining for $E_L = 1$. Poisson's ratios for rocks and linings must be assumed when determining these stiffness matrices. The finite element mesh must be chosen in such a way that the measuring points coincide with the nodes of the mesh.

Considering Equation (8.4) and the equivalent nodal forces given by Equation (8.1) into Equation (8.3), we obtain

$$[K^*]\{u\} = \sigma_{x0}/E_R\{P_1\} + \sigma_{y0}/E_R\{P_2\} + \tau_{xy0}/E_R\{P_3\} \tag{8.5}$$

where $\{P_i\}$ $(i = 1\text{–}3)$ denotes the equivalent nodal forces, corresponding to the components of unit initial stress.

Substituting $\sigma_{x0}/E_R = l$ and $\sigma_{y0}/E_R = \tau_{xy0}/E_R = 0$ into Equation (8.5) gives

$$[K^*]\{u\} = \{P_1\} \tag{8.6}$$

Solving Equation (8.6), we obtain the displacement $\{u_x\}$ at nodal points due to the initial stress component, $\sigma_{x0}/E_R = l$ only. By following a similar procedure, the displacements $\{u_y\}$, $\{u_{xy}\}$ due to the other components of unit initial stress $\sigma_{y0}/E_R = l$ and $\tau_{xy0}/E_R = l$, respectively, are obtained.

Considering the displacements $\{u_x\}$, $\{u_y\}$ and $\{u_{xy}\}$ due to each component of the initial stress, the following equation is derived:

$$[A]\{\sigma_0\} = \{u\} \tag{8.7}$$

where $[A] = [\{u_x\}, \{u_y\}, \{u_{xy}\}]$

$$\{\sigma_0\} = \{\sigma_{x0}/E_R, \quad \sigma_{y0}/E_R, \quad \tau_{xy0}/E_R\}$$

$\{\sigma_0\}$ is called the "normalised initial stress".

The displacement $\{u\}$ may be split into two parts, i.e. measured displacements $\{u_1\}$ and unknown displacements $\{u_2\}$ at the nodes.

$$[A_1]\{\sigma_0\} = \{u_1\} \tag{8.8}$$

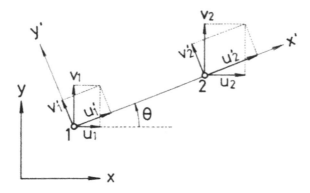

Figure 8.1   Relative displacements between adjacent measuring points.

The matrix $[A_1]$ is uniquely defined when Poisson's ratio for the ground and lining materials and the parameter R are given. It is noted that the matrix $[A_1]$ becomes the same as that derived by the stiffness matrix method (Sakurai & Takeuchi, 1983), while the flexibility matrix method can significantly reduce computation time.

It should be pointed out that in engineering practice it is easier to measure relative displacements between two different measuring points than absolute displacements; hence $\{u_1\}$ in Equation (8.8) should be transformed to relative displacements. The relative displacements between adjacent measuring points shown in Figure 8.1 can be expressed in terms of absolute displacements as follows:

$$\{\Delta u\} = \begin{Bmatrix} u'_2 - u'_1 \\ v'_2 - v'_1 \end{Bmatrix} = \begin{bmatrix} -\cos\theta - \sin\theta & \cos\theta & \sin\theta \\ \sin\theta - \cos\theta & -\sin\theta & \cos\theta \end{bmatrix} \begin{Bmatrix} u_1 \\ v_1 \\ u_2 \\ v_2 \end{Bmatrix} \tag{8.9}$$

Hence the absolute displacements $\{u_1\}$ are related to the measured relative displacements $\{\Delta u_m\}$ by

$$\{\Delta u_m\} = [T]\{u_1\} \tag{8.10}$$

The transformation matrix $[T]$ is obtained with the consideration of Equation (8.9).

Substituting Equation (8.8) into Equation (8.10) yields

$$[A_1^*]\{\sigma_0\} = \{\Delta u_m\} \tag{8.11}$$

with

$$[A_1^*] = [T][A_1]$$

If the least squares method is adopted in Equation (8.11), the normalised initial stresses can be determined uniquely from the measured relative displacements by

$$\{\sigma_0\} = [[A^*]^T[A^*]]^{-1}[A^*]^T\{\Delta u_m\} = [F]\{\Delta u_m\} \tag{8.12}$$

When determining the normalised initial stress, the displacements at all the nodal points can be calculated by Equation (8.7). The strain in each element, therefore, can be obtained by using the following relationship between strains and displacements at nodes:

$$\{\varepsilon\} = [B]\{u\} \tag{8.13}$$

where the matrix $[B]$ is the nodal point location function.

The above-mentioned back analysis procedure was formulated by the Finite Element Method (FEM), and its computer program Direct Back Analysis Program (DBAP) was developed at Kobe University. DBAP has been applied to various tunnel construction projects. Two examples of the use of DBAP are shown in the following Section 8.3.

8.3 CASE STUDY I (WASHUZAN TUNNELS)

8.3.1 Exploration tunnel (work tunnel)

8.3.1.1 Introduction

Two double-track railway tunnels and two double-lane highway tunnels were being constructed in a relatively homogeneous geological formation consisting of weathered granite. Although the length of the tunnels is only about 200 m, this project had particular features in that the four tunnels are situated very close to each other, and the overburden is only about 30 m (see Figure 8.2). In order to obtain the input data for the design of the main tunnels, a work tunnel for carrying construction machines, construction materials and muck debris was constructed in advance of the excavation of the main tunnels. The work tunnel was located almost parallel to the main tunnels, as shown in Figure 8.2.

8.3.1.2 Displacement measurements and back analyses

Field measurements were carried out around the work tunnel during its construction. The displacements due to the excavation were measured by using Sliding Micrometer-ISETH developed in ETH, Zurich (Kovari & Amstad, 1983) together with an inclinometer, which were inserted from the ground surface. The convergences of displacements of the inner surface of the tunnel were also measured, as shown in Figure 8.3. All the measured displacements were used as input data for the back analysis procedure, which is described in Section 8.2. As a result, the normalised initial stress was calculated by Equation (8.12), resulting in Young's modulus and initial stress of rock masses around the tunnels being determined, if needed (Sakurai & Takeuchi, 1983).

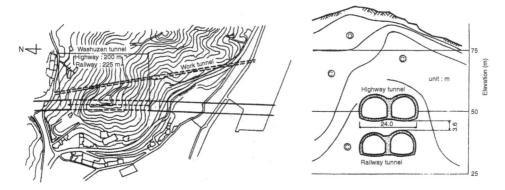

Figure 8.2 Plane view of tunnel construction site and tunnel cross section with geological condition.

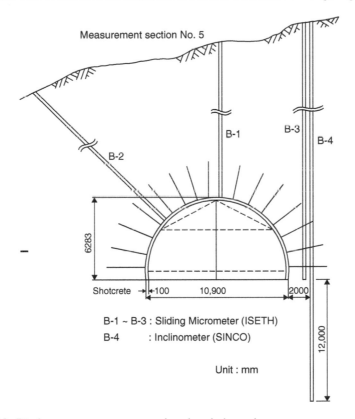

Figure 8.3 Displacement measurements along boreholes and convergence measurements.

Back analyses are carried out for determining the normalised initial stress, where Poisson's ratio is assumed as $v = 0.3$. The back-calculated normalised initial stress is shown as follows (tensile stress is positive):

$$\sigma_{x0}/E = -7.79 \times 10^{-4}, \quad \sigma_{y0}/E = -8.00 \times 10^{-4}, \quad \tau_{xy0}/E = -2.48 \times 10^{-4}$$

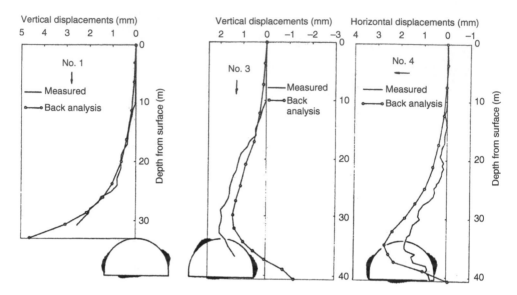

Figure 8.4 Comparison of back-calculated displacements with measured values.

If we assume the vertical component of initial stress as being approximately equal to the overburden pressure, i.e. $\sigma_{y0} = -0.95$ MPa, then the other components of initial stress as well as Young's modulus can be uniquely determined as follows:

$$\sigma_{x0} = -0.93 \text{ MPa}, \quad \tau_{xy0} = -0.30 \text{ MPa}, \quad E = 1190 \text{ MPa}$$

In this back analysis the effect of shotcrete was disregarded.

In order to verify the accuracy of the back analysis, we calculate the displacements by using both the back-calculated initial stress and Young's modulus in the ordinal (forward) finite element analysis, and the back-calculated displacements are compared with the measured displacements. Some of the results are shown in Figure 8.4. It can be seen that the back-calculated displacements show a fairly good agreement with the measured displacements.

8.3.1.3   Design analysis of the main tunnels

Since the initial stresses and Young's modulus have been determined by the back analyses carried out at the work tunnel, they are used as input data for the design of the main tunnels in such a way that the ordinary (forward) finite element analyses are performed for designing support measures together with the construction procedures. In order to assess the stability of the tunnels, the maximum shear strain distributions around the main tunnels are calculated in each excavation step. The finite element mesh for the main tunnels is shown in Figure 8.5. One of the results is illustrated in Figure 8.6, which indicates the maximum shear strain distribution around the main tunnels at all the excavations of the tunnels being completed.

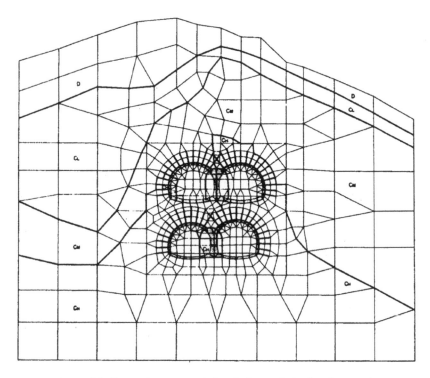

Figure 8.5 Finite element mesh for design analysis of main tunnel.

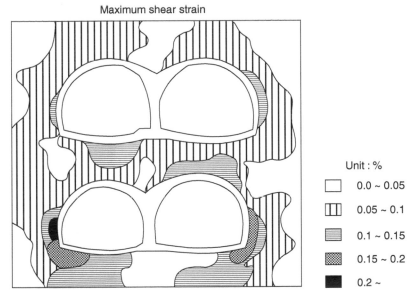

Figure 8.6 Maximum shear strain distribution around the main tunnels at all the excavations of the tunnels being completed.

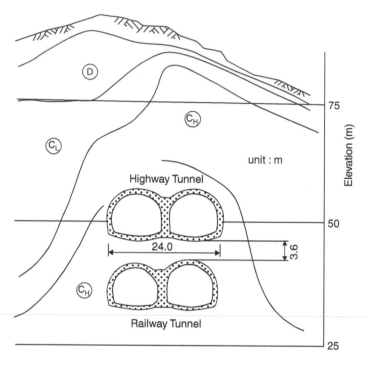

Figure 8.7 Tunnel cross section with geological formations.

For assessing the stability of the tunnels, the maximum shear strains occurring around the tunnels are compared with the critical shear strain of the geomaterials (Sakurai et al., 1993a).

8.3.2 Excavation of the main tunnels

8.3.2.1 Brief description with respect to the tunnels and instrumentation

Two double-lane highways and two double-track railway tunnels were constructed at a shallow depth. They consisted of two upper and two lower tunnels, parallel and adjacent to each other. The ground in which the tunnels were bored consisted of highly fractured and weathered granite. An example of the cross section of the four tunnels together with the geological formation is shown in Figure 8.7.

An intensive field measurement system has been carefully planned and executed. In this measurement system, displacement measurements are highlighted, where the sliding micrometers and "Trivec" developed by Kovari and his colleagues (Kovari & Amstad, 1983; Koeppel et al., 1983) are extensively used. High-precision inclinometers are also employed. One of the measurement sections is given in Figure 8.8.

The tunnel excavation was begun from the lower west railway tunnel, followed by the lower east. After completion of the two railway tunnels with a reinforced concrete lining, two pilot tunnels for the upper highway tunnels were bored in advance, and then the main portion of the upper tunnels was excavated.

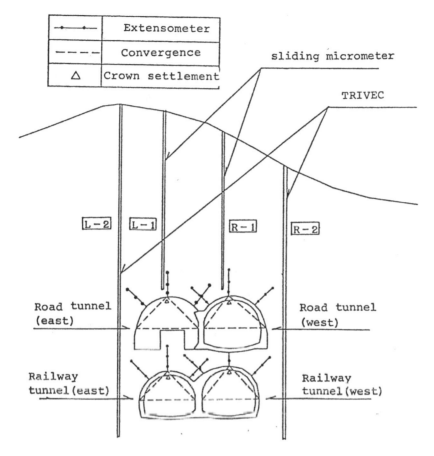

Figure 8.8 One of the measurement sections.

8.3.2.2 Back analysis of measured displacements

The systematic field measurements and back analyses were performed throughout the excavation. The normalised initial stress was back-calculated from measured displacements, thereby enabling Young's modulus to be determined if necessary. As an example, one set of back analysis results is shown in Table 8.1. It should be noted that the values of the normalised initial stress vary for each excavation step. This is simply because stress in the ground around tunnels is influenced by excavations, as well as by the fact that initial stress differs from place to place.

It is of interest to know that Young's modulus slightly decreases with the progress of the excavations until the main excavation of the upper highway tunnels has begun. This may be due to the fact that the ground surrounding the tunnels tends to be loosened by the excavations and, thus, large displacements occur in the tunnels excavated afterwards. In the excavation of the upper highway tunnels, however, a group of horizontal pipes was installed near the tunnel crown prior to the excavation in order to

Table 8.1 Normalised initial stress and Young's modulus determined by back analysis (tensile stress is positive).

Excavation steps	Normalised initial stress			Young's modulus (MPa)
	σ_x/E	σ_y/E	τ_{xy}/E	
West railway tunnel*	-0.199×10^{-2}	-0.111×10^{-2}	0.126×10^{-3}	536
East railway tunnel*	-0.140×10^{-2}	-0.153×10^{-2}	-0.554×10^{-3}	389
Pilot tunnel**	-0.236×10^{-2}	-0.956×10^{-3}	0.899×10^{-4}	330
West highway tunnel**	-0.819×10^{-3}	-0.302×10^{-3}	0.700×10^{-4}	761
East highway tunnel**	-0.477×10^{-3}	-0.262×10^{-3}	0.171×10^{-4}	1136

*: Back-calculated from sliding micrometer and TRIVEC measurement results
**: Back-calculated from crown settlement and convergence measurement results

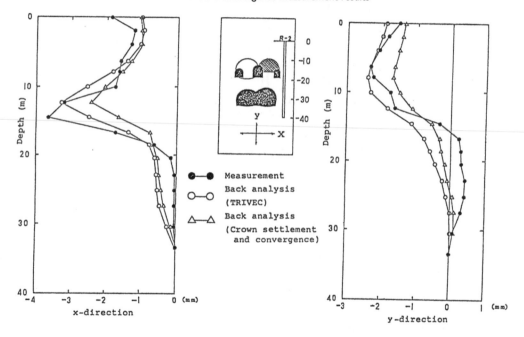

Figure 8.9 Both vertical and horizontal displacements obtained from TRIVEC measurements and back analysis (due to the excavation of the upper half of the west highway tunnel). In the back analysis two cases being calculated by using (a) TRIVEC measurements alone, and (b) crown settlement and convergence alone.

stabilise the ground. This may give a large value of Young's modulus. It should also be noted that Young's modulus obtained by the back analysis described here is not necessarily the mechanical constant of the material itself, but that of the material systems with reinforcement, such as rock bolts and other support structures. The normalised initial stress is then used as input data for the ordinary (forward) finite element analyses to calculate displacements and strains. As an example, we show the results of this calculation performed during the excavation of the upper highway tunnels. Some of the displacement results are shown in Figures 8.9 and 8.10, which are compared with the measured displacements.

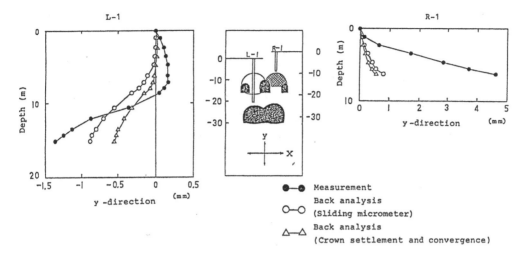

Figure 8.10 Vertical displacements obtained from the sliding micrometer measurements and back analysis (due to the excavation of the upper half of the west highway tunnel). In the back analysis two cases being calculated by using (a) the sliding micrometer measurements alone, and (b) crown settlement and convergence alone.

It is seen from Figure 8.9 that the back-calculated displacements approximately coincide with the measured displacements. However, in Figure 8.10 there lies a large discrepancy between the two. This discrepancy indicates that a loosening zone may exist above the tunnel crown, which cannot be simulated by the back analysis program (DBAP) used here, because it is formulated for homogeneous isotropic linear elastic materials. To reduce the discrepancy between the back-calculated displacements and the measured displacements, non-elastic strains should be considered in the back analysis, which is described in Chapter 9 as the "universal back analysis method".

8.3.2.3 *Assessment of the stability of tunnels*

The stability of tunnels can be assessed in such a way that the maximum shear strains occurring around the tunnels are compared with the critical shear strain of the geomaterials (Sakurai et al., 1993a). If the maximum shear strain is still smaller than the critical shear strain, then the tunnels are stable.

The maximum shear strains can be determined by back analysis of measured displacements in the following numerical procedure. The displacements around a tunnel are determined by back analysis of measured displacements as shown in the previous section, and strain distributions around the tunnel can be calculated by Equation (8.13). Once the strain distributions around the tunnel are determined, the maximum shear strain γ_{max} can be easily calculated by Equation (8.14).

$$\gamma_{max} = |\varepsilon_1 - \varepsilon_3| \tag{8.14}$$

where ε_1: maximum principal strain
ε_2: minimum principal strain

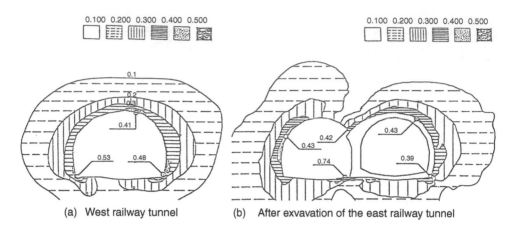

Figure 8.11 Maximum shear strain (%) around tunnels.

As already described in Sections 8.1 and 8.2, if only the displacements are needed, it is not necessary to split the normalised initial stress into two values, i.e. the initial stress and Young's modulus. The normalised initial stress alone can provide the displacements around tunnels in the use of Equation (8.7) without knowing both the initial stress and Young's modulus of the ground materials. This is a great advantage for the use of normalised initial stresses in back analysis, because it is not necessary to know the initial stress as well as Young's modulus in determining strain distribution around tunnels.

In this case study, the displacement distributions around the tunnels have already been determined by back analyses of measured displacements. The maximum shear strain γ_{max} can be easily calculated by Equation (8.14).

As an example, some of the results of the maximum shear strain distributions around the tunnels are shown in Figures 8.11 and 8.12. It is seen from these figures that the value of the maximum shear strain is less than 1% throughout the region around the tunnels. In order to assess the stability of tunnels, these maximum shear strains back-calculated from measured displacements are compared with the critical shear strain of the geomaterials (Sakurai et al., 1993a), or compared with the hazard warning levels shown in Figure 7.5.

8.4 CASE STUDY II (TWO-LANE ROAD TUNNEL IN SHALLOW DEPTH)

8.4.1 Introduction

This case study deals with the interpretation of displacement measurement results obtained during the excavation of a two-lane road tunnel. In this project, DBAP is used for investigating the deformational behaviour of the tunnel (Noami et al., 1987).

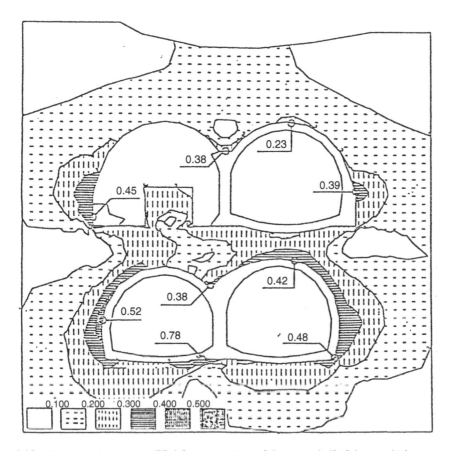

Figure 8.12 Maximum shear strain (%) (after excavation of the upper half of the east highway tunnel).

Since the back analysis program has been developed by assuming the ground consists of homogeneous isotropic elastic materials, if the back-calculated displacements around the tunnel approximately coincide with the measured displacements, then the deformational behaviour of the ground is more or less homogeneous isotropic elastic materials. If some discrepancy lies between the two, some non-elastic deformational behaviour of the ground tends to occur. To establish clearly what kind of non-elastic deformational mechanism starts to occur, the non-elastic strain should be considered in the back analysis together with the normalised initial stress (Sakurai et al., 1994a).

8.4.2 Brief description of the tunnel

The total length of the tunnel is 126 m. The height of the maximum overburden is 25 m, while some sections have an overburden of only 1.5–10 m. Thus, the tunnel can be classified as very shallow. The ground consists of weathered granite which becomes a nearly soil-like material. The velocity of elastic wave (Vp) travelling through the ground is approximately 0.8–1.0 km/sec. Little underground water comes out at the tunnel surface, and its magnitude is about 20–30 l/min at some portions of the tunnel. The excavation of the tunnel was conducted by the New Austrian Tunnelling method (NATM), starting from the top heading followed by the lower half approximately 30 m behind the face of the top heading. A permanent arch lining was installed 30 m behind the face of the lower half, and invert concrete was laid after completion of the arch lining. Support measures consist of the combination of shotcrete, rock bolts and steel ribs, depending on the geological conditions. The standard tunnel cross section with support measures is shown in Figure 8.13.

The longitudinal section of the tunnel is shown in Figure 8.14, where geological conditions with rock classifications are indicated. The measuring sections are also shown with measuring instruments in the figure.

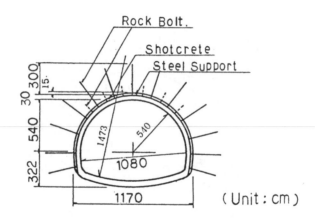

Figure 8.13 Standard tunnel cross section.

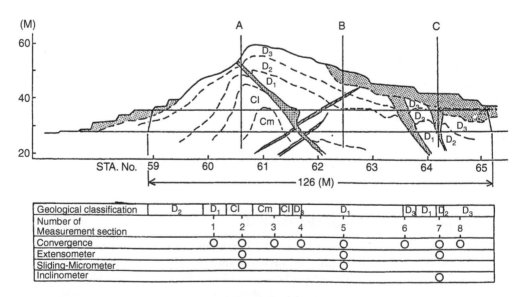

Figure 8.14 Location of measurement section.

8.4.3   Field measurements

Field measurements were performed at eight different tunnel sections, as shown in Figure 8.14, where displacements of the ground, axial force of rock bolts, and stress acting in shotcrete were measured. Convergence metres, multi-rod extensometers and inclinometers were used to measure the displacements around the tunnel, and a sliding micrometer developed at ISETH, Zurich was installed from the ground surface. The settlement at the tunnel crown was also measured. The places where the instruments were installed for taking the displacement measurements are shown in Figure 8.15. The results of the displacement measurements are shown in the following section.

8.4.3.1   Convergence measurements

The final values of convergence at each measurement section are illustrated in Figure 8.16. It should be noted that the largest value obtained is either for measuring line No. 1 (convergence), or for the settlement at the tunnel crown. The figure also indicates that large displacements appear at Section Nos. 6, 7 and 8, where the height of the overburden is as small as less than 10 m. In order to secure the stability of the tunnel excavated in the ground with a small overburden, additional rock bolts and shotcrete were installed. The invert at the top heading of Section No. 8 was temporarily covered with shotcrete to increase the stability of the tunnel.

8.4.3.2   Multi-rod extensometer and sliding micrometer measurements

The multi-rod extensometer and sliding micrometer were used to measure the displacements caused by the excavation at Sections A and B. The results are shown

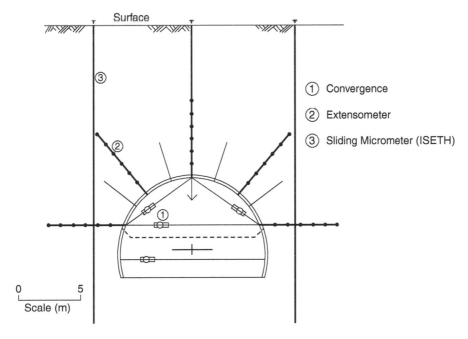

Figure 8.15 Installation of instruments (sections A and B).

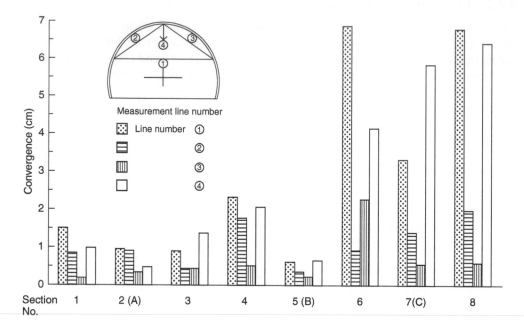

Figure 8.16 Final value of convergence at different measurement sections.

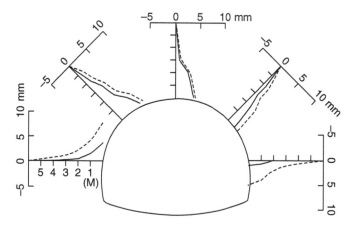

Figure 8.17 Results of multi-rod extensometer measurements (Section A).

in part, in Figures 8.17 and 8.18, indicating the displacements measured by using the multi-rod extensometers and the sliding micrometer, respectively. In these figures, the displacements measured before and after excavation of the lower half are compared.

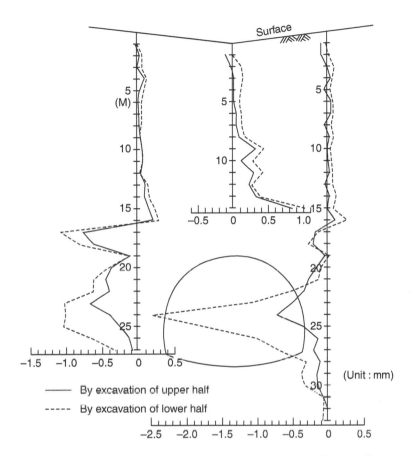

Figure 8.18 Results of sliding micrometer measurements (Section A).

8.4.4 Back analysis of measured displacements

Back analyses were carried out for determining strain distributions around the tunnel from measured displacements. In the back analysis, four cases are considered as input data depending on the different type of instruments, that is, Case 1: convergence measurements alone, Case 2: both convergence and extensometer measurements, Case 3: multi-rod extensometer measurements alone, and Case 4: both extensometer and sliding micrometer measurements.

It should be noted that the sliding micrometer can measure total displacements of the ground due to the excavations, while the others measure only the partial displacements which take place after the measurements start. Therefore, in Case 4 only the incremental displacements measured after the tunnel face passed through the measuring section are used as input data for the back analyses, resulting in an easy comparison of back-calculated displacements with the measured values being achieved. The results of the back analyses conducted for Section B are summarised in Table 8.2.

Assuming that the vertical initial stress is equal to the overburden pressure, we can determine each component of initial stresses and Young's modulus of the geomaterials

Table 8.2 Results of back analyses (Section B).

	Case 1 Convergence	Case 2 Convergence extensometer	Case 3 Multi-rod extensometer	Case 4 Extensometer sliding micrometer
σ_{x0}/E	-0.0977×10^{-3}	-0.123×10^{-2}	-0.221×10^{-2}	-0.233×10^{-2}
σ_{y0}/E	-0.669×10^{-3}	-0.978×10^{-3}	-0.178×10^{-2}	-0.233×10^{-2}
τ_{xy0}/E	-0.243×10^{-3}	0.957×10^{-4}	0.245×10^{-3}	0.206×10^{-3}
σ_{x0} (MPa)	-0.686	-0.592	-0.585	-0.470
σ_{y0} (MPa)	-0.470	-0.470	-0.470	-0.470
τ_{xy0} (MPa)	-0.170	-0.046	0.065	0.041
E (MPa)	703.2	481.0	264.9	201.6
ν (assumed)	0.3	0.3	0.3	0.3

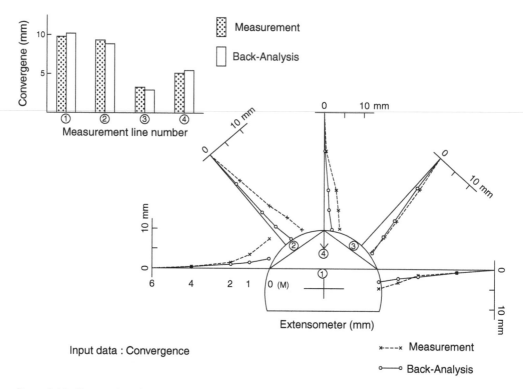

Input data : Convergence

Figure 8.19 Comparison between measured and back-calculated displacements (Case 1, input data: convergences alone).

from the normalised initial stress obtained by back analyses of measured displacements, as shown in Table 8.2.

Once both the initial stresses and Young's modulus are determined, they are used as input data for the ordinary (forward) finite element analysis to determine stress, strain and displacement distributions in the ground. The comparison between calculated (back-calculated) displacements and measured displacements is shown in Figure 8.19

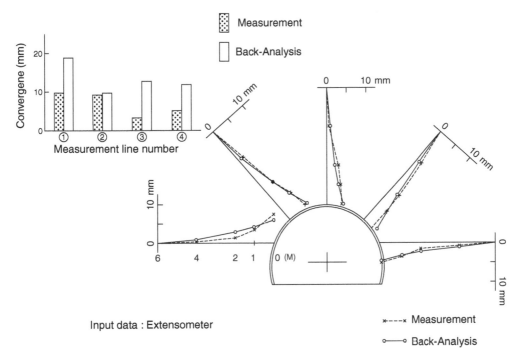

Figure 8.20 Comparison between measured and back-calculated displacements (Case 3, input data: multi-rod extensometer measurements alone).

(Case 1) and Figure 8.20 (Case 3). In Figure 8.19 only the convergence measurement results are used as input data. The figure shows that there lies a large discrepancy between the measured and back-calculated displacements appearing in the extensometer measurements, while a good agreement can be seen for the convergences. It is no wonder that only convergence measurement results are used as input data in back analyses, while the extensometer measurement results are not considered in the back analyses.

As seen in Figure 8.19, the displacements back-calculated by using the convergence measurements alone become smaller than the displacements measured by multi-rod extensometers. This implies that the displacements near the tunnel inner surface are small, even though large displacements occur in a large extent of rock masses around the tunnel.

On the other hand, for Case 3 shown in Figure 8.20, a good agreement exists between the displacements measured by multi-rod extensometers and back-calculated displacements, while the convergence shows a large discrepancy between back-calculated displacements and the measured values. This is simply because of the fact that the convergence results are not considered in the back analyses as input data.

Looking at the results of Case 1 together with Case 3, we can easily imagine that Case 2 may give some of the discrepancy between back-calculated displacements and the measured values for both convergence and extensometer measurements. It is noted that, considering the back analysis results of both Case 1 and Case 3, the deformational mechanism of the tunnel is such that the ground around the tunnel may be loosened due to excavations, followed by a concrete lining being placed on the inner surface of the tunnel. As a result, the loosened zone contracts as tunnel excavations progress.

The displacements back-calculated by using the results of both the multi-rod extensometer measurements and sliding micrometer measurements (Case 4) are compared with the measured displacements, as shown in Figure 8.21. It is seen from the figure that there is a good agreement between the back-calculated displacements and measured values. This means that a large extent of the ground around the tunnel deforms as a homogeneous isotropic linear elastic material. It is of interest to know that in Case 4 the back-calculated Young's modulus becomes the smallest among all the cases, as shown in Table 8.2. This means that, if we use the data of the measured displacements over a large extent of the ground, the back-calculated Young's modulus of the ground tends to be a smaller value.

Considering the results for both Case 1 and Case 2, however, we understand that special care must be paid to the deformational behaviour of the ground behind

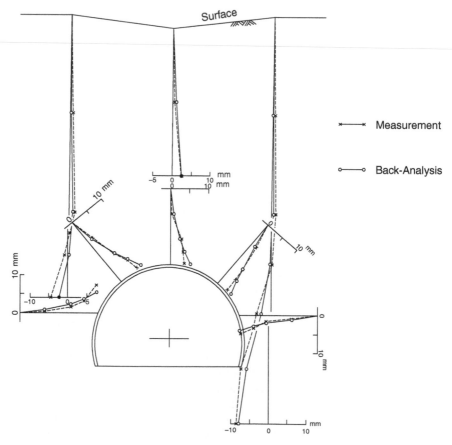

Figure 8.21 Comparison between measured and back-calculated displacements (Case 4, input data: both multi-rod extensometer and sliding micrometer measurements).

the tunnel lining, because the convergence measurement results do not coincide with the extensometer measurement results, in such a way that small convergences are measured, while large displacements are measured by the multi-rod extensometers, as shown in Figure 8.19. This may be due to the fact that the tunnel was reinforced by support measures such as rock bolts, shotcrete, steel ribs and concrete linings, resulting in a reduction in the convergence, as shown in Figure 8.19. This means that the support measures form a ground arch around the tunnel, which can reduce the displacements in the ground reinforced by the support measures. To simulate this behaviour, non-elastic strains such as plastic, creep, loosening, contracting, etc. occurring in the near-field around the tunnel should be taken into consideration in addition to the normalised initial stress. In other words, both the normalised initial stress and non-elastic strains can be back-calculated at the same time from measured displacements (Sakurai et al., 1994a). The back analysis method is briefly described in Section 9.2.

8.4.5   Assessment of the stability of tunnels

In order to assess the stability of tunnels, the maximum shear strains occurring around the tunnels are compared with the critical shear strain of the geomaterials (Sakurai et al., 1993a). The maximum shear strains can be determined by back analyses of measured displacements, as described in the previous section. For this case study, one of the results (Case 3: extensometer measurements) is shown in Figure 8.22, which indicates the contour lines of maximum shear strain distributions.

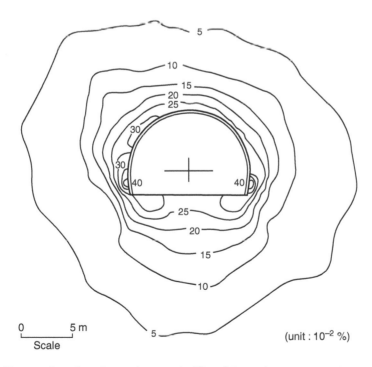

Figure 8.22 Contour line of maximum shear strain (Case 3, input data: extensometer measurements).

This figure shows that the largest value of the maximum shear strain reaches approximately 0.4%, ranging from 0.2 to 0.3% in the ground at a distance of 5 m from the tunnel inner surface where the rock bolts were installed. Considering the reinforcements by support measures, the maximum shear strain distributions shown in Figure 8.22 should be modified considering the non-elastic strain in the near-field around the tunnel, because, as already described, the ground near the tunnel inner surface may be loosened due to excavations, followed by support structures such as rock bolts and shotcrete being placed on the inner surface of the tunnel, resulting in the loosened zone being contracted as tunnel excavations progress.

Chapter 9

Universal back analysis method

9.1 INTRODUCTION

The universal back analysis method (non-elastic strain approach) has been developed for assessing the stability of tunnels considering non-elastic strains in the back analyses. Since the details of the back analysis method have been published elsewhere (Sakurai et al., 1993b, 1994a), only a brief description is given in the next section, Section 9.2. According to this method, both the normalised initial stress and non-elastic strains can be back-calculated at the same time from measured displacements.

The back analysis method aims to achieve the rational design and construction of tunnels excavated in the ground having non-linear mechanical characteristics. The method should be capable of interpreting the results of displacement measurements obtained during the excavations, in such a way that the maximum shear strains back-calculated considering the results of measured displacements occurring around tunnels are compared with the critical shear strain of the geomaterials, and if the maximum shear strains are smaller than the critical shear strain, the tunnels are stable, enabling the excavations to proceed. If, however, the maximum shear strains tend to be greater that the critical shear strain, additional support measures must be placed. Therefore, the back analysis method must be capable of determining the maximum shear strain distributions around the tunnels by taking into account the non-elastic strains occurring around the tunnels.

9.2 MATHEMATICAL FORMULATION CONSIDERING NON-ELASTIC STRAIN

The total strain occurring in the ground due to excavation can generally be divided into two parts, that is, elastic and non-elastic, as given in the following equation:

$$\{\varepsilon\} = \{\varepsilon_e\} + \{\varepsilon_0\} \tag{9.1}$$

where $\{\varepsilon\}$: total strain
$\{\varepsilon_e\}$: elastic strain
$\{\varepsilon_0\}$: non-elastic strain

In the non-elastic part, all the non-elastic strains such as those caused by plastic characteristics of geomaterials, loosening due to gravitational force, and blasting, contraction, creep, etc. are included. Hence, Hooke's law is written as:

$$\{\sigma\} = [D]\{\varepsilon_e\}$$
$$= [D](\{\varepsilon\} - \{\varepsilon_0\}) \qquad (9.2)$$

Taking into account Equation (9.2) in the formulation of finite element analysis, the following equation is derived:

$$[K]\{u\} = \{P\} + \{P_0\} \qquad (9.3)$$

where $[K]$: stiffness matrix
$\quad$ $\{u\}$: displacement vector at nodal points

$$\{P\} = \int_{v1} [B]^T\{\sigma_0\}dV - \int_{v1} [N]^T\{b\}dV \qquad (9.4)$$

$$\{P_0\} = -\int_{v2} [B]^T [D] \{\varepsilon_0\}dV \qquad (9.5)$$

where $\{P\}$: nodal forces corresponding to the excavation
$\quad$ $\{\sigma_0\}$: initial stress
$\quad$ $\{b\}$: gravitation force
$\quad$ $\{P_0\}$: external nodal force caused by a non-elastic strain
$\quad$ $[K]$: stiffness matrix for isotropic linear elastic materials
$\quad$ $[B]$: matrix containing the derivatives of the displacement interpolation function
$\quad$ $[D]$: elastic constitutive matrix.

In tunnelling practice, displacement measurements are most commonly carried out by using convergence meters and borehole extensometers. Therefore, in finite element analysis some of the displacements at the nodal points are known. Considering these known displacements in Equation (9.3), the non-elastic strain can be back-calculated. In this back analysis procedure any optimisation program can be used. If the number of measurement data points is greater than the number of unknown parameters, a least squares method can be used. However, in the proposed back analysis method all the components of non-elastic strains become unknown parameters, resulting in the number of unknown parameters usually being far greater than the number of measurement data points. Therefore, back analyses with a large number of unknown parameters cannot in general be performed. This is a great disadvantage for the back analysis method. To overcome this disadvantage, some sort of constraint is introduced to obtain a reliable solution.

As an example of reducing the number of unknown parameters, a simple way is to assume that a certain extent around the tunnel is divided into a few regions, and each region has a constant distribution of non-elastic strains. This assumption can reduce the number of unknowns so that a least squares method can be used. A case study of back analysis carried out by introducing a constraint around tunnels is shown in Section 9.3.

Nevertheless, the back analysis procedure considering non-elastic strains is greatly advantageous, because there is no need to assume any mechanical models for the back analysis; hence it can cover any type of non-linear deformational mechanism. With regard to non-elastic strains, we can perform back analyses in such a way that the discrepancy between measured and calculated displacements is minimised by changing the non-elastic strains. This is the reason why the back analysis procedure is named "universal back analysis".

Let us now give an example to demonstrate the applicability of the non-elastic strain approach to engineering practice. The example problem is the loosening of jointed rock masses which often occurs around tunnels due to blasting. The mechanism of the rock masses must be volume expansion, and it can be considered a non-elastic strain. In the same manner as shown in Equation (9.5), the non-elastic volumetric strain can be easily taken into account in the back analysis, by replacing Equation (9.5) by the following Equation (9.6).

$$\{P_0\} = -\int_{v2} [B]^T [D] \{\varepsilon_{0v}\} dV \tag{9.6}$$

where $\{\varepsilon_{ov}\}$ is non-elastic volumetric strain, which can be back-calculated from the measured displacements. Non-elastic volumetric strain $\{\varepsilon_{ov}\}$ is expressed in two-dimensional plane strain state, as follows:

$$\{\varepsilon_{0v}\} = \left\{ \begin{array}{c} \varepsilon_{x0v} \\ \varepsilon_{y0v} \\ \gamma_{xy0v} \end{array} \right\} = \frac{e_0}{3} \left\{ \begin{array}{c} 1+v \\ 1+v \\ 0 \end{array} \right\} \tag{9.7}$$

where e_0 is non-elastic volumetric strain defined as follows:

$$e_0 = \varepsilon_{x0v} + \varepsilon_{y0v} + \varepsilon_{z0v} \tag{9.8}$$

v: Poisson's ratio

Contini et al. (2007) proposed the volume strain for simulating the expansion of the ground due to grouting. The nodal forces are given in terms of the increment form as:

$$\{\Delta f_g\} = \sum \int_{v} [B]^T [D] \{\Delta \varepsilon_{vol}^{ne}\} dV \tag{9.9}$$

where $\{\Delta f_g\}$ is the increment of the nodal forces

$\{\Delta\varepsilon_{vol}^{ne}\}$ is the increment of strain vector corresponding to the imposed volume deformation due to grouting

$\{\Delta\varepsilon_{vol}^{ne}\}$ can be back-calculated by minimising the error function given in Equation (2.3) for the discrepancy between measured and computed displacements.

It is noted that each element has a non-elastic volumetric strain $\{\varepsilon_{0v}\}$, whose unknown parameter is one, i.e. e_0 alone. This means that, in the case of back analyses for the volume expansion and/or contraction, the number of unknown parameters is reduced to one-third of the case of all three components of non-elastic strain $\{\varepsilon_0\}$ being back-calculated. Furthermore, the volume expansion and/or contraction may often occur only near tunnel inner surfaces. The expansion is caused by blasting, while the contraction is due to the stiff linings being installed. This means that the number of unknown parameters to be back-calculated is usually not too large, which is a definite advantage of the universal back analysis method.

9.3 CASE STUDY (TUNNEL EXCAVATED IN SHALLOW DEPTH)

9.3.1 Tunnel configuration and instruments

A case study is shown for demonstrating the applicability of the back analysis procedure considering non-elastic strains in practice. In this case study, a double-track railway tunnel located in a shallow depth was constructed in a densely populated urban area. It is excavated in fine-grained sand deposits of the diluvial formation. The overburden height is about $H = 0.8\,D$ (D: tunnel diameter). Both the tunnel diameter and the height of the overburden are approximately 10 m (Sakurai, 1996).

As support structure, shotcrete reinforced by steel ribs was placed on the inner surface of the tunnel. The cross section of the tunnel together with the shotcrete and steel ribs support structure is shown in Figure 9.1. The tunnel was excavated by means of NATM with six excavation steps, as shown in the numbering in the figure.

Extensometers and inclinometers were installed from the ground surface before the tunnel excavation so that the total displacements due to the excavation could be measured. The location of extensometers and inclinometers is shown in Figure 9.2. Surface settlements were also measured by ordinary surveying of levelling.

9.3.2 Back analyses

Back analyses were carried out for determining the initial stresses, Young's modulus and non-elastic strains from the measured displacements. The finite element mesh used in this back analysis is shown in Figure 9.3.

In this figure the regions numbering ①–⑧ indicate those in which an identical number of non-elastic strains is assumed to occur, resulting in the number of unknown parameters (non-elastic strains) becoming smaller than the number of measurement data points. Therefore, a least squares method can be applied, and the back analysis results are given in Table 9.1.

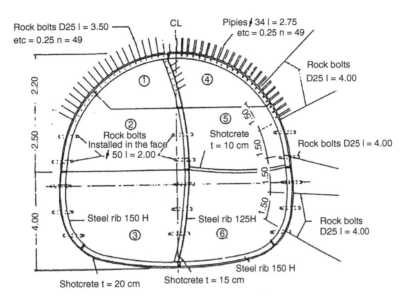

Figure 9.1 Tunnel cross section together with support structures.

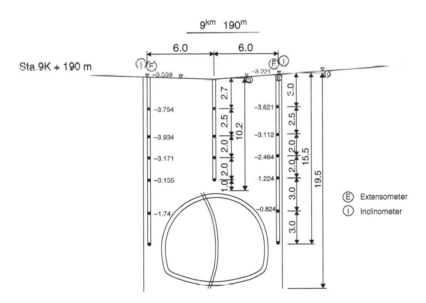

Figure 9.2 Arrangement of measurement devices.

Displacements can then be calculated by an ordinary (forward) finite element method using all the back analysed data shown in Table 9.1. The calculated displacements are compared with the measured values. The results are shown in Figures 9.4 and 9.5. It is obvious from these figures that there is a good agreement between the two.

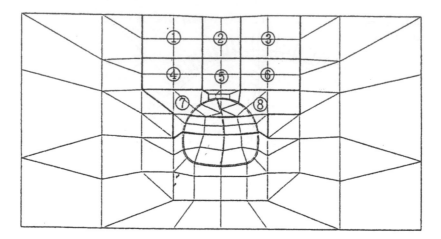

Figure 9.3 Finite element mesh.

Table 9.1 Results of back analyses.

	σ_{0x}	σ_{0y}	τ_{0xy}	E
Initial stress (MPa)	−0.0858	−0.2880	0.0027	
Elastic modulus (MPa)				104.4

	Zones	ε_x	ε_y	γ_{xy}
Non-elastic strain (%)	1	−1.983	0.022	−0.026
	2	0.098	0.006	−0.028
	3	−1.367	−0.003	0.375
	4	−1.779	0.115	−1.202
	5	1.779	0.115	1.480
	6	−1.471	−0.078	0.603
	7	−0.541	0.095	−1.267
	8	−0.679	−0.201	0.178

The maximum shear strain distribution is shown in Figure 9.6, which indicates a loosening zone existing above the tunnel arch. The maximum shear strain is then compared with an allowable strain for assessing the stability of the tunnel. The critical shear strain proposed by Sakurai et al. (1993a) can be used as a criterion for assessing the stability of tunnels.

In the back analysis, if we assume the mechanical model of the ground consists of homogeneous and isotropic elastic material, it is obvious that no loosening zone exists (see Figure 3.3). This is not surprising because the loosening of materials has not been taken into account in the back analysis. As already mentioned, in back analysis the mechanical model of rock masses should not be assumed, but it should be taken into account in the analysis without assuming it. To achieve this requirement, the non-elastic strain approach must be a powerful tool, resulting in no assumption being

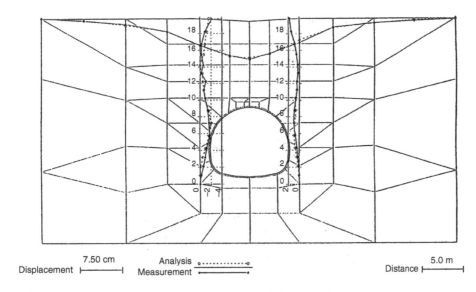

Displacement ⊢————⊣ 7.50 cm   Analysis ⊙·············⊙   Measurement •————•   Distance ⊢————⊣ 5.0 m

Figure 9.4 Comparison between measurements and analysis results (surface settlements and horizontal displacements).

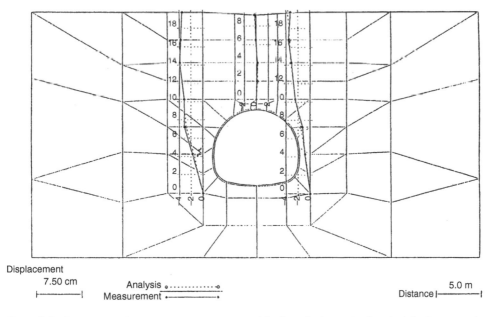

Displacement
⊢————⊣ 7.50 cm   Analysis ⊙·············⊙   Measurement •————•   Distance ⊢————⊣ 5.0 m

Figure 9.5 Comparison between measurements and back analysis results (vertical displacements).

made in the modelling of rocks; rather, a computer can find a non-elastic behaviour, such as loosening, from the results of the measured displacements. (Note: Figure 9.6 is identical to Figure 3.4 which is compared with Figure 3.3. Both figures show the results of back analyses carried out using identical input data of measured displacements.)

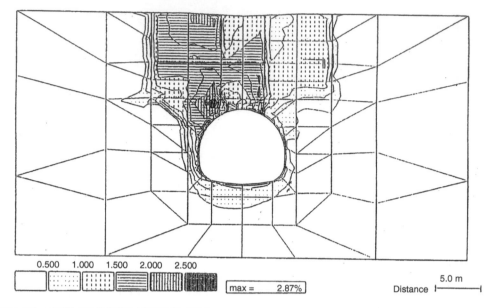

0.500 1.000 1.500 2.000 2.500

max =   2.87%

5.0 m

Distance ⊢————⊣

Figure 9.6 Maximum shear strain distribution (non-elastic strain approach).

In back analysis, the mechanical model of materials is assumed in general, and never modified during calculation to minimise the error function. Hence, this calculation procedure should be called parameter identification, which should differ from back analysis. In parameter identification, only the mechanical parameters are identified, while the mechanical model remains unchanged during the calculation. It must be obvious that parameter identification does not work well in rock engineering practice.

9.3.3 Supporting mechanism of rock bolts, shotcrete and steel ribs

In the case study described here, rock bolts were installed inward from the tunnel surface, and shotcrete and steel ribs were placed on the tunnel inner surface. However, in back analysis such support structures are simply modelled as a single tunnel lining with the equivalent value of Young's modulus, which can be determined so as to minimise the error function given in Equation (9.10), which indicates the degree of agreement between the measured and the computed displacements.

$$\delta = \frac{\sum\limits_{i=1}^{N} (u_i^m - u_i^c)^2}{\sum\limits_{i=1}^{N} u_i^m} \to \min \tag{9.10}$$

where u_i^m and u_i^c are the measured and computed displacements at the measuring point i, respectively. N is the total number of measurements.

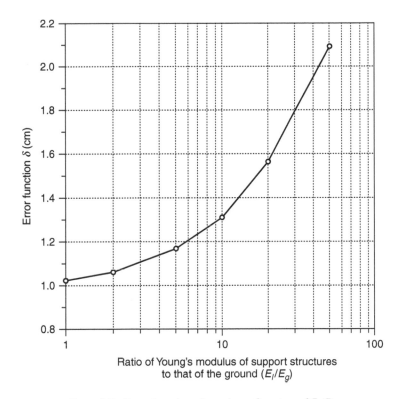

Figure 9.7 Error function plotted as a function of E_l/E_g.

The error function defined by Equation (9.10) is plotted as a function of the ratio of the equivalent Young's modulus of the support structures E_l to that of the ground materials E_g, as shown in Figure 9.7.

It is interesting to see from the figure that the best agreement between measured and computed displacements by back analysis were obtained for the case of $E_l/E_g = 1.0$. This means that the equivalent Young's modulus of the support structures is the same as that of the ground. In other words, the best agreement is obtained in the case of the tunnel being unlined. It is a surprising result that the tunnel behaves just like an unlined tunnel, even though the tunnel is reinforced with stiff support structures such as rock bolts, shotcrete and steel ribs. However, it should be noted that the apparent Young's modulus of the ground clearly increases with the support structures. This case study demonstrates the difficulty in modelling support structures such as rock bolts, shotcrete and steel ribs, and that misleading conclusions can easily be derived if an improper model is adopted.

It is also noted that tunnel support structures, such as shotcrete, rock bolts, steel ribs and concrete lining, do not support the earth pressure directly, but cause the rigidity of the ground to increase. This is the exact concept behind the mechanism of supporting structures installed in tunnels constructed by NATM.

9.4 MODELLING OF SUPPORT STRUCTURES

9.4.1 Modelling of rock bolts

Sakurai (2010) proposed a design approach for rock bolts stabilising tunnels, emphasising that the rock bolts should not be modelled separately from the ground, but should be modelled together with the ground by considering the interaction mechanism between rock bolts and rock joints. In this design approach, back analyses play a major role in evaluating the rock masses reinforced by rock bolts as overall tunnel supporting systems.

Mechanical parameters, such as Young's modulus and the strength parameters of rock masses, can be determined by *in situ* plate bearing tests and direct shear tests, respectively. It is noted that the mechanical parameters determined by these *in situ* tests are those for a continuum equivalent to the rock masses containing various joints. In other words, the rock masses concerned are assumed implicitly to be a continuous material in which all the joints disappear. Consequently, if rock bolts are installed in such a continuous "jointed rock mass", the rock bolts end up being modelled as if they were installed in a continuum, as shown in Figure 9.8(a). This modelling procedure may result in the underestimation of the effect of rock bolts reinforcing jointed rock masses.

It is obvious that the values of both Young's modulus and the compressive strength of jointed rocks, particularly hard rocks, increase greatly by the installation of rock bolts, due to the fact that the relative displacements along joints are constrained by the rock bolts. However, if rock bolts are installed in the equivalent continuum, the

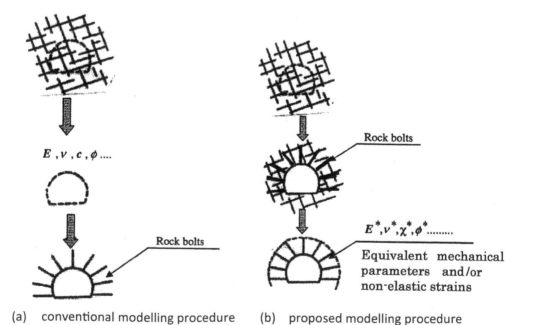

(a) conventional modelling procedure   (b) proposed modelling procedure

Figure 9.8 Modelling procedure for jointed rock masses reinforced by rock bolts (Sakurai, 2010).

constraining effect of the rock bolts on the joints cannot be properly taken into account. This is a shortcoming of the continuum approach to modelling jointed rock masses reinforced with rock bolts.

In the numerical analyses, all the joint systems in rock masses could be modelled explicitly by using discrete element methods. However, it is almost impossible to detect all the joint systems, and the mechanical characteristics of filling materials of joints have hardly been evaluated. Furthermore, the interaction mechanism between rock bolts and the ground materials are also extremely complex. To overcome such difficulties in modelling of jointed rock masses, the continuum approach must be preferable in engineering practice, that is, the jointed rock masses reinforced with rock bolts should be modelled such that both rock bolts and the ground are modelled together considering the interaction between them, as shown in Figure 9.8(b).

It is noted that the mechanical parameters of the equivalent continuum reinforced with rock bolts can be evaluated in such a way that they are determined locally in each region reinforced with rock bolts considering their orientation in relationship to joint direction. In jointed rock masses reinforced with rock bolts, both the joints and the rock bolts should be modelled together by considering the interaction mechanism between them. As a result, the mechanical parameters must differ from place to place depending on the orientation of rock bolts, because the orientation of rock bolts differs along the tunnel inner surface, although the joint systems are homogeneous. The mechanical parameters in each region reinforced with rock bolts can be determined considering the effect of rock bolts constraining joint movements.

The question may now arise in engineering practice of how to evaluate the mechanical parameters of jointed rock masses reinforced with rock bolts. To answer this question, back analyses considering non-elastic strains, described in Section 9.2, must be used for determining the mechanical parameters from measured displacements. It is noted that the effect of interaction between rock bolts and rock masses is implicitly included in the maximum shear strain distributions around tunnels determined by the back analyses of measured displacements.

On the other hand, in the case of soft rocks, whose mechanical behaviour may be mainly due to the matrix of rocks, not joints, the effect of the rock bolts constraining the movement of joints is less compared with that of hard rocks, resulting in only a small difference between the conventional approach and the proposed one. Hence, it may be possible to model soft rocks reinforced with rock bolts by the conventional method, as shown in Figure 9.8(a).

9.4.2 Modelling of shotcrete and steel ribs

The modelling of shotcrete with steel ribs placed on the inner surface of tunnels is similar to the case of rock bolts, that is, shotcrete should not be modelled separately from the surrounding rock masses, but should be modelled together with rock masses, considering the interaction mechanism between the shotcrete and rock masses.

Chapter 10

Initial stress of rock masses determined by boundary element method

10.1 INTRODUCTION

In the design of underground openings, such as tunnels, shafts, underground caverns, etc., both initial stresses and mechanical properties of rock masses must first be determined by *in situ* investigation. To measure the initial stress, overcoring stress relief methods and hydrofracturing tests are well known. The initial stresses determined by these methods are ones for a limited local area. However, rock masses over a large extent are in general non-homogeneous materials containing various kinds of joint systems, resulting in the initial stresses varying from place to place. Therefore, in the design of large-scale underground structures, such as underground hydropower plants, the initial stresses necessary for the design of the structures as input data must be the average value of the large extent of rock masses. This means that the overcoring stress relief methods and hydrofracturing tests cannot be used, because they measure the initial stresses at the pinpoint of rock masses. To overcome this problem, the number of measuring points for the overcoring stress relief methods and hydrofracturing tests can be increased. However, this causes technical and financial problems. In addition, underground hydropower plants usually have a complex configuration, so that three-dimensional initial stresses must be determined.

A promising way to determine the average value of initial stress over a large extent of rock masses is to use a back analysis procedure. In the back analyses, the field measurement results are needed as input data for the analyses, which are obtained during excavation of exploration tunnels. Displacement measurements using convergence meters and multi-rod borehole extensometers are preferable because of the simplicity in instrumentation, and the reliability of the data obtained. The initial stresses and material properties can then be determined by back analysis of these measured displacements.

The back analysis procedure can be used for determining the normalised initial stress, which is defined as the ratio of initial stresses to Young's modulus (Sakurai & Takeuchi, 1983; Sakurai & Shinji, 1984). It is noted that the normalised initial stress alone is sufficient to determine the maximum shear strain distributions around the tunnels. In order to assess the stability of tunnels, the maximum shear strains can be compared with the hazard warning levels, shown in Figure 7.5. As already described, the maximum shear strain distribution is uniquely determined from the normalised initial stress. This means that it is not necessary to know the initial stress and Young's modulus separately from each other. If necessary, however, the values of Young's modulus and initial stresses can be derived from the normalised initial stresses by assuming the vertical component of initial stresses being equal to the overburden pressure, as already described earlier.

In the case of a large underground hydropower plant, its configuration is generally very complex; hence the back analysis procedure should be formulated in three-dimensional ways. For the two-dimensional cases, a back analysis program (DBAP) has already been developed as described in Section 8.2, formulated by the Finite Element Method (FEM).

In this chapter, three-dimensional back analysis is described, which is based on the idea of normalised initial stress. In order to reduce the number of elements, the three-dimensional Boundary Element Method (BEM) is used for the mathematical formulation of a computer program. A case study for demonstrating the applicability of the proposed back analysis program based on BEM is presented, and it shows that the method was applied successfully to back-calculate the three-dimensional normalised initial stress, enabling the three-dimensional initial stresses and Young's modulus of a large extent of rock masses to be determined.

There are two different types of mathematical formulation of BEM. One is the direct method and the other is the indirect method (Shimizu & Sakurai, 1983). In the following section, the direct method of three-dimensional BEM is described together with a case study applied to an underground hydropower plant (Sakurai & Shimizu, 1986).

10.2 THREE-DIMENSIONAL BACK ANALYSIS METHOD

10.2.1 Mathematical formulation of the method

The back analysis method is formulated by a three-dimensional boundary element method. Since a detailed formulation of this back analysis is presented elsewhere, only a brief description is given here.

In the formulation of the back analysis, the following assumptions are made:

1   The mechanical behaviour of the ground is idealised by an isotropic linear elastic model, so that the material constants reduce to Young's modulus and Poisson's ratio only. Since Poisson's ratio has less influence on the results of the analysis, an appropriate value can be used.
2   The initial state of stress is assumed to be constant all over the region under consideration and compressive stress is taken as positive.

Displacements due to excavation at a point p in the ground (see Figure 10.1) are derived from Somigliana's identity, as follows:

$$u_i(p) = \int_S U_{ki}(q,p)t_k(q)dS_q - \int_S T_{ki}(q,p)u_k(q)dS_q \tag{10.1}$$

where $U_{ki}(q,p)$ is the well-known Kelvin solution, corresponding to a concentrated force acting at the point p in the infinite elastic space.

$$U_{ij}(q,p) = \frac{1+v}{8\pi E(1-v)r}\{(3-4v)\delta_{ij} + r_{,i}\,r_{,j}\} \tag{10.2}$$

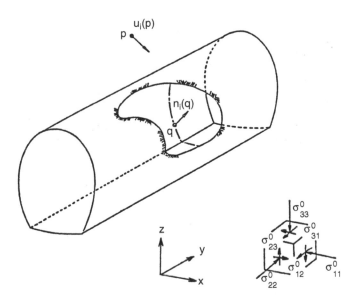

Figure 10.1 Displacements due to excavation of an underground opening.

where δ_{ij} is Kronecker's delta, r is the distance between points p and q, and E and n denote Young's modulus and Poisson's ratio, respectively. $T_{ki}(q,p)$ is the fundamental solution of traction corresponding to the Kelvin solution. The traction vector $t_i(q)$ is given as follows:

$$t_i(q) = -n_j(q)\sigma_{ji}^0 \tag{10.3}$$

where σ_{ji}^0 represents the initial stress in the ground, and $n_i(q)$ denotes the normal unit vector at point q on the surface of the underground opening. The following linear relationship between displacement $\{u\}$ and the initial stress can be derived from Equations (10.1)–(10.3):

$$\{u\} = [A]\{\sigma_0\} \tag{10.4}$$

where $\{\sigma_0\}$ is defined as:

$$\{\sigma_0\} = \{\sigma_x^0/E \quad \sigma_y^0/E \quad \sigma_z^0/E \quad \tau_{yz}^0/E \quad \tau_{zx}^0/E \quad \tau_{xy}^0/E\}^T \tag{10.5}$$

$\{\sigma_0\}$ is called the normalised initial stress, as already defined earlier in a two-dimensional state. E denotes Young's modulus of the ground. It should be noted that matrix $[A]$ is a function of Poisson's ratio of rock masses, the shape of the underground openings, and the location at and the direction in which the displacements are measured. Knowing the location of the extensometers, therefore, the matrix is uniquely defined for a given Poisson's ratio.

Equation (10.4) consists of the same number of equations as the number of displacement measurement data points and contains six unknown values of the normalised initial stress. If the number of measurements is six, Equation (10.4) gives a simultaneous equation to solve the normalised initial stress. If the number is greater than six, the normalised initial stress can be determined by an optimisation procedure. When adopting the least squares method, Equation (10.4) yields:

$$\{\sigma_0\} = ([A]^T[A])^{-1}[A]^T\{u_m\} \qquad (10.6)$$

where $\{u_m\}$ is a vector of the measured displacements. This equation gives the normalised initial stress uniquely determined from the measured displacements for a given Poisson's ratio. It is noted that in Equation (10.4) the measurements of the relative displacement between two measurement points as well as absolute displacements can be used as input data (Gioda & Jurina, 1981; Sakurai & Takeuchi, 1983). It follows that even simple convergence measurements alone carried out at the surface of the underground openings are sufficient for determining a three-dimensional initial stress state. In addition, the three-dimensional sequence of steps for excavation and the period for installation of measuring instruments can easily be taken into account in this back analysis. The components of initial stress and Young's modulus can be separated from the normalised initial stress by assuming that the vertical stress is equal to the overburden pressure, i.e.:

$$\sigma_z^0 = \gamma H \qquad (10.7)$$

where H and γ denote the overburden at the datum point and the specific weight of the ground respectively.

10.2.2   Computational stability

Measured displacements always show some scattering due to a variety of reasons, e.g. complex characteristics of rock masses, heterogeneous geological conditions, measurement errors, etc. This means that, in general, a back analysis method in geomechanics should derive a numerically stable solution for any scattered measurement data. In order to assure the computational stability of the method, a numerical simulation has been conducted, and it has been verified that the method is stable enough to provide sufficiently accurate results even for scattered input data.

10.3   CASE STUDY

To demonstrate the applicability of the above-mentioned back analysis method for engineering practice, a case study of an underground powerhouse is shown in the following (Sakurai & Shimizu, 1986).

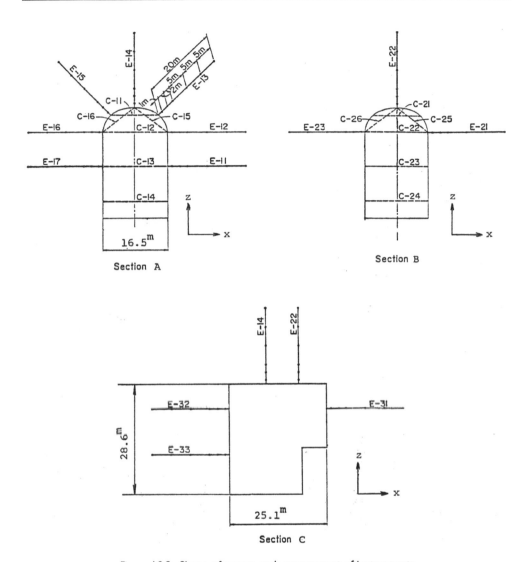

Figure 10.2 Shape of cavern and arrangement of instruments.

The ground in which the powerhouse was built consists of fresh and solid coarse-grained biotite granite. The powerhouse is located with an overburden of about 200 m. The cavern is 28.6 m high, 16.5 m wide and 25.1 m long, as shown in Figure 10.2. The arrangement of the measurements is also given in this figure.

The back analysis was conducted using the measured displacements taken at the completion of the fourth excavation stage, Lift 4 (see Figure 10.3). The assumption of

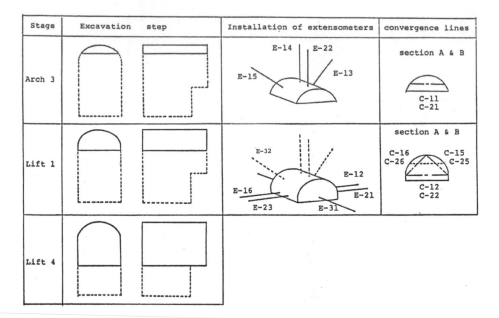

Figure 10.3 Installation of measuring instruments and sequence of excavation steps.

$v = 0.3$ leads immediately to the normalised initial stress.

$$\{\sigma_0\} = \begin{Bmatrix} \sigma_x^0/E \\ \sigma_y^0/E \\ \sigma_z^0/E \\ \tau_{yz}^0/E \\ \tau_{zx}^0/E \\ \tau_{xy}^0/E \end{Bmatrix} = \begin{Bmatrix} 0.72 \times 10^{-3} \\ 0.90 \times 10^{-3} \\ 0.40 \times 10^{-3} \\ -0.65 \times 10^{-3} \\ -0.15 \times 10^{-3} \\ -0.69 \times 10^{-3} \end{Bmatrix}$$

The normalised initial stress obtained may now be split into the initial stress and Young's modulus by assuming the vertical component of initial stress.

$$\sigma_z^0 = 23.5(\text{kN/m}^3) \times 200(\text{m}) = 4.70 \, \text{MPa}$$

and

$$E = 11.75 \, \text{GPa}$$
$$\sigma_x^0/E = 8.46 \, \text{MPa}$$
$$\sigma_y^0/E = 10.58 \, \text{MPa}$$
$$\tau_{yz}^0/E = -7.64 \, \text{MPa}$$
$$\tau_{zx}^0/E = -1.76 \, \text{MPa}$$
$$\tau_{xy}^0/E = -8.11 \, \text{MPa}$$

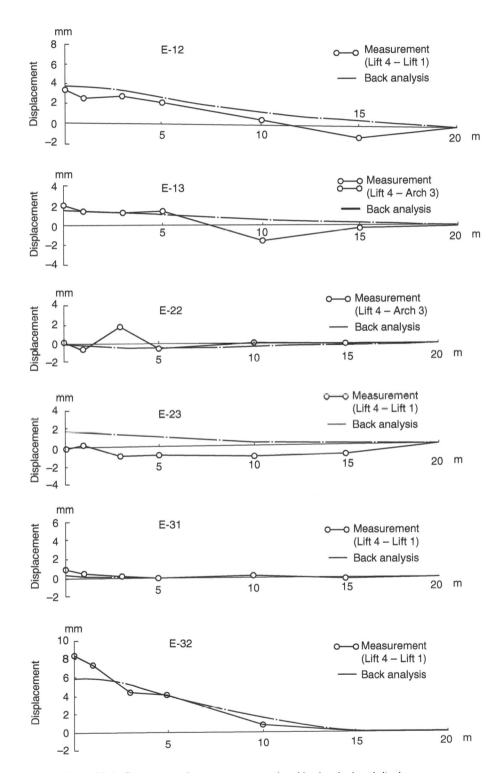

Figure 10.4 Comparison between measured and back calculated displacements.

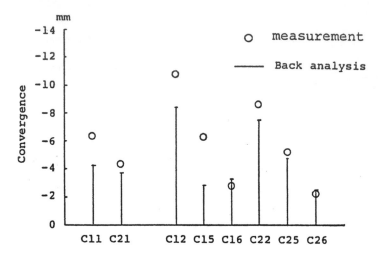

Figure 10.5 Comparison between measured and back calculated convergences.

In conclusion, the back analysis method described here is based on the three-dimensional boundary element method, so that the sequence of steps for excavation and the period for installation of measuring instruments can easily be considered in the back analysis.

Chapter 11

Back analysis for the plastic zone occurring around underground openings

11.1 INTRODUCTION

In order to assess the stability of underground openings, it is of primary importance to evaluate the extent of the plastic zone occurring around the openings. In principle, if all the material constants for elasto-plastic characteristics of soils/rocks as well as the initial stress of the ground are known, the plastic zone can be calculated by an ordinary elasto-plastic stress analysis. However, an ordinary stress analysis may not be an effective means to evaluate the plastic zone, because it is extremely difficult to know all such information about the ground. Until now, there has been no reliable method available to determine the plastic zone.

In this chapter, a back analysis method for determining the plastic zone appearing around underground openings is presented. Of course, if the material constants and initial stress are back-calculated, the plastic zone can then be calculated by an ordinary elasto-plastic analysis. In general, however, a back analysis for determining the material constants of cohesion and the internal friction angle is not an easy task because of the non-linearity of the problems.

Sakurai et al. (1985) proposed a simple back analysis method for determining the plastic zone, which is based on an interpretation of the maximum shearing strain obtained by the measured displacements. The procedure of the method is as follows:

First, the normalised initial stress (initial stress divided by Young's modulus) is back-calculated from the measured displacements. In this back analysis, only the linear elastic finite element analysis is sufficient because the ground is assumed to be an equivalent linear elastic material.

Second, an ordinary linear elastic finite element analysis is carried out using the normalised initial stress as its input data, and the maximum shear strain distribution is obtained. Finally, the maximum shear strain occurring around the underground openings is compared with the critical shear strain of soils and rocks. One of the contour lines of the maximum shear strain then gives the boundary between the elastic and plastic zones. In this method, however, evaluation of the critical shear strain is in question and is a most difficult task.

In the following, a method for evaluating the critical shear strain is described and the applicability of this method is demonstrated by means of a computer simulation.

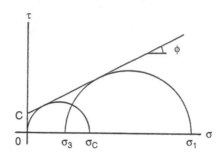

Figure 11.1 Mohr-Coulomb's yielding criterion.

11.2 ASSUMPTIONS

The mechanical properties and initial stress of the ground are assumed as follows:

1   The ground consists of homogeneous isotropic elastic and perfectly plastic material which conforms to the Mohr-Coulomb yield criterion (see Figure 11.1).
2   Hooke's law is assumed for the stress-strain relationship in the elastic zone and it is supposed that no volume change occurs in the plastic zone.
3   The initial stress existing in the ground prior to excavation is constant in the area of excavation (see Figure 11.2).
4   A plane strain condition is assumed in the computer simulation. The compressive stress is taken as positive here.

11.3 FUNDAMENTAL EQUATIONS

11.3.1 Maximum shear strain on the elasto-plastic boundary

The Mohr-Coulomb criterion is expressed as:

$$\sigma_1 - \sigma_3 = (\sigma_1 + \sigma_3)\sin\phi + 2c\cos\phi \tag{11.1}$$

where σ_1 and σ_3 ($\sigma_1 > \sigma_3$) are principal stresses, and c and ϕ are cohesion and the internal friction angle, respectively. The principal stress on the elasto-plastic boundary must satisfy Hooke's law. Hence, the following equation can be derived on the boundary:

$$\varepsilon_1 - \varepsilon_3 = \frac{1+\nu}{E}(\sigma_1 - \sigma_3) \tag{11.2}$$

where ε_1 and ε_3 ($\varepsilon_1 > \varepsilon_3$) are principal strains, and E and ν denote Young's modulus and Poisson's ratio, respectively. From Equations (11.1) and (11.2), the maximum shear strain on the elasto-plastic boundary is expressed as:

$$\varepsilon_1 - \varepsilon_3 = \frac{1+\nu}{E}(\sigma_1 + \sigma_3)\sin\phi + 2\frac{1+\nu}{E}c\cos\phi \tag{11.3}$$

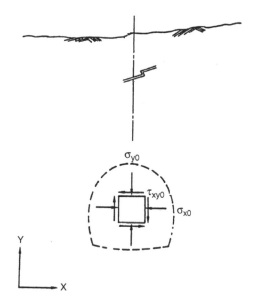

Figure 11.2 Initial stress uniformly distributed throughout rocks being excavated.

Equation (11.3) is rewritten as follows by introducing the uniaxial critical strain ε_0, $\varepsilon_0 = \sigma_c/E = 2c\cos\phi/E(1 - \sin\phi)$ (σ_c: uniaxial compressive strength), proposed by Sakurai (1981).

$$\gamma_c = 2(1 + v)\frac{\bar{p}}{E}\sin\phi + (1 - \sin\phi)(1 + v)\varepsilon_0 \qquad (11.4)$$

where $\bar{p} = (\sigma_1 + \sigma_3)/2$ is the average value of the maximum and minimum principal stresses on the elasto-plastic boundary. γ_c defined by Equation (11.4) is the critical shear strain.

11.3.2 Relationship between real and equivalent Young's modulus

In order to determine the critical shear strain, it is seen from Equation (11.4) that Young's modulus E, Poisson's ratio v, internal friction angle ϕ and critical strain ε_0 must be known in advance. The values of v, ϕ and ε_0 can be easily determined from laboratory test results. The most difficult task in evaluating γ_c in Equation (11.4) is to determine Young's modulus E of rocks, particularly of the jointed rock masses. One of the possibilities for determining E is to back-calculate it from the deformational behaviour of the tunnels observed during excavation. If a linear elastic model is used in the back analysis, Young's modulus can easily be obtained.

This elastic back analysis does not give the real value of Young's modulus, however, if a plastic zone appears around openings. The value determined from the elastic back analysis is called an equivalent Young's modulus and usually differs from the real value

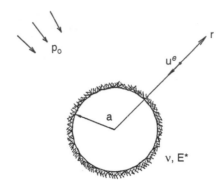

Figure 11.3 Circular tunnel excavated in elastic ground under hydrostatic initial stress.

needed for Equation (11.4). Therefore, in the case of the existence of a plastic zone, the real Young's modulus must be determined. For this purpose, we should know the relationship between real Young's modulus E and equivalent Young's modulus E^*, so that E can be derived from E^*. The relationship between E and E^* can be obtained as explained in the following.

The radial displacement around an unlined circular tunnel excavated in an elastic ground which is under a hydrostatic initial state of stress is expressed as:

$$u^e = \frac{1+v}{E^*} p_0 \frac{a^2}{r} \tag{11.5}$$

where p_0 is the initial stress (see Figure 11.3).

On the other hand, the radial displacement around an unlined circular tunnel excavated in the elasto-plastic ground under hydrostatic initial stress is expressed as (see Figure 11.4):

$$u^p = \frac{1+v}{E} (p_0 \sin\phi + c \cos\phi) \left(\frac{\lambda}{a}\right)^2 \frac{a^2}{r} \tag{11.6}$$

where

$$\frac{\lambda}{a} = \left\{ (1-\sin\phi)\left(\frac{p_0}{c}\tan\phi + 1\right)\right\}^{\frac{1-\sin\phi}{2\sin\phi}}$$

$$c = \varepsilon_0 E(1-\sin\phi)/2\cos\phi$$

where λ is the radius of the elasto-plastic boundary.

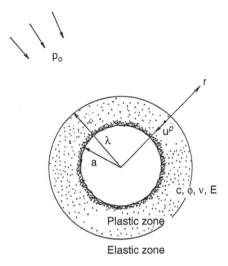

Figure 11.4 Circular tunnel excavated in elasto-plastic ground under hydrostatic initial stress.

Equating Equations (11.5) and (11.6) gives the relationship between E and E^* as follows:

$$E = \frac{E^* \sin \phi}{\left[\left(\dfrac{2p_0}{\varepsilon_0 E} - 1\right) \sin \phi + 1\right]^{-\frac{1-\sin\phi}{\sin\phi}} - \dfrac{\varepsilon_0 E^*}{2p_0}(1 - \sin \phi)} \tag{11.7}$$

The real value of Young's modulus is then calculated by Equation (11.7) and substituted into Equation (11.4) to determine the critical shear strain.

11.4 THE METHOD FOR DETERMINING THE ELASTO-PLASTIC BOUNDARY

The procedure for determining the elasto-plastic boundary from the measured displacements is as follows:

1 Displacements around openings due to excavation are measured by convergence and borehole extensometer.
2 The normalised initial stress is back-calculated from the measured displacements. Poisson's ratio is assumed. The method of back analysis described in Chapter 8 can be adopted for this purpose.
3 Ordinary elastic finite element analysis is conducted through use of the normalised initial stress to determine the maximum shear strain occurring due to the excavation.

4   The initial strain $\{\varepsilon_0\}$ is calculated by the following equation:

$$\{\varepsilon_0\} = \left\{ \begin{array}{c} \varepsilon_x^0 \\ \varepsilon_y^0 \\ \gamma_{xy}^0 \end{array} \right\} = \begin{bmatrix} 1-\nu & -\nu & 0 \\ -2\nu & 1 & 0 \\ 0 & 0 & 2(1+\nu) \end{bmatrix} \{\bar{\sigma}_0\} \tag{11.8}$$

where $\{\sigma_0\} = \{\sigma_{xo}/E^* \quad \sigma_{yo}/E^* \quad \tau_{xy0}/E^*\}^T$ is the normalised initial stresses. The total maximum shear strain is then calculated as:

$$\gamma_{\max} = \sqrt{\left\{ \left(\bar{\varepsilon}_x^0 + \Delta\varepsilon_x \right) - \left(\bar{\varepsilon}_y^0 + \Delta\varepsilon_y \right) \right\}^2 + \left(\bar{\gamma}_{xy}^0 + \Delta\gamma_{xy} \right)^2} \tag{11.9}$$

where $\Delta\varepsilon_x$, $\Delta\varepsilon_y$ and $\Delta\gamma_{xy}$ are components of strain due to excavation.

5   The critical shear strain of soils/rocks is calculated by Equation (11.4) in which $\overline{P}$ is the maximum principal stress component of the initial stress. The real Young's modulus which is necessary for this calculation is obtained by Equation (11.7) in which p_0 is the maximum principal stress component of the initial stress. The initial stress components are determined from the normalised initial stress by assuming the vertical stress component of initial stress to be the same as the overburden pressure. The internal friction angle and uniaxial critical strain are assumed by considering laboratory experiment results.

6   The boundary between the elastic and plastic zones can be obtained as a contour line of the total maximum shear strain being the same value as the critical shear strain. If the critical shear strain is greater than any of the total maximum shear strains occurring around the openings, no plastic zone exists.

11.5 COMPUTER SIMULATION

11.5.1 Procedure

In order to demonstrate the applicability of the method described above, a computer simulation is conducted. In the simulation, the displacements around an opening due to excavation are first calculated by an ordinary elasto-plastic finite element method, and they are used as virtual "measured displacements" in the computer simulation. In order to verify the accuracy of the method, the elasto-plastic boundary obtained by the proposed back analysis is compared with the one calculated by the ordinary elasto-plastic finite element analysis.

11.5.2 An example problem and simulation results

A horseshoe-shaped unlined tunnel with a radius of 5 m is considered here as an example (see Figure 11.5). The initial state of stress is non-hydrostatic and the magnitudes of principal stress are $p_1 = 3.92$ MPa and $p_2 = 1.96$ MPa. Their directions are shown in Figure 11.5. The mechanical constants are listed in Table 11.1.

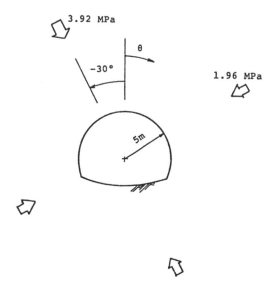

Figure 11.5 A horseshoe-shaped unlined tunnel under non-hydrostatic initial stress.

Table 11.1 Mechanical constants used in the simulation.

Young's modulus	E (MPa)	196
Poisson's ratio	v	0.3
Cohesion	c	0.49
Internal friction angle	ϕ (%)	30
Critical strain	ε_0 (MPa)	0.87

The elasto-plastic finite element analysis is first conducted to obtain the "measured displacements". The extent of the plastic zone obtained by the analysis is shown in Figure 11.6 as a shaded area. The back analysis of the "measured displacements" provides the following normalised initial stress:

$$\{\sigma_0\} = \begin{Bmatrix} 2.17 \times 10^{-2} \\ 2.69 \times 10^{-2} \\ 0.62 \times 10^{-2} \end{Bmatrix} \tag{11.10}$$

Some of the contour lines of the total maximum shear strain are shown in Figure 11.6. The critical shear strain in this case is $\gamma_c = 3.01\%$. The contour line corresponding to this value is shown by a dotted line in this figure, which gives the outer boundary of the plastic zone. It is seen that there is a fairly good agreement between the back-calculated results and those calculated by elasto-plastic finite element analysis.

The results of this back analysis are summarised in Table 11.2.

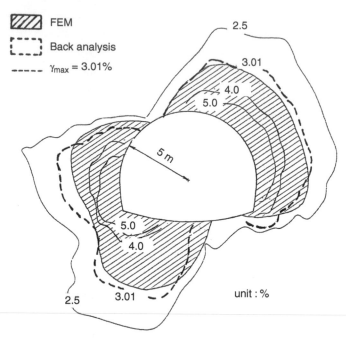

Figure 11.6 Plastic zone evaluated by the back analysis method.

Table 11.2 Results of the back analysis.

	Back analysis results	Exact value
σ_1^0 (MPA)	3.96	3.92
σ_3^0 (MPA)	2.25	1.96
θ (deg.)	−33.7	−30
E^* (MPa)	128	–
E (MPa)	210	196
γ_c (%)	3.01	–
c (MPa)	0.53	0.49

In conclusion, the computer simulation has demonstrated that the back analysis method described here is highly accurate for determining the boundary between the elastic and plastic zones. The method requires only a linear elastic finite element analysis, so that it is possible to perform this back analysis at the construction site to monitor the plastic zone occurring around the openings.

Chapter 12

Back analysis considering anisotropy of rocks

12.1 INTRODUCTION

In this chapter a back analysis method is proposed for determining displacement (strain) distributions around the underground openings excavated in jointed rock masses. The deformational behaviour of jointed rock masses may be classified into the following three modes: (1) spalling of joints, (2) sliding along a particular slip surface, and (3) plastic flow. Consequently, the mechanical model used in the back analyses must represent all these deformational modes. In the proposed back analysis methods, a simple linear mechanical model is introduced, in which the three different deformational modes can be analysed by using a single model based on continuum mechanics.

12.2 CONSTITUTIVE EQUATIONS

In order to simulate all three modes of deformation, i.e. spalling, sliding, and plastic flow, the following constitutive equation is proposed (Sakurai & Ine, 1986).

The proposed equation is expressed in terms of $x'-y'$ local coordinate systems as (see Figure 12.1):

$$\{\sigma'\} = [D']\{\varepsilon'\} \tag{12.1}$$

where

$$[D'] = \frac{E_2}{(1 + v_1)(1 - v_1 - 2nv_2^2)} \begin{bmatrix} n(1 - nv_2^2) & nv_2(1 + v_1) & 0 \\ nv_2(1 + v_1) & 1 - v_1^2 & 0 \\ 0 & 0 & m(1 + v_1)(1 - v_1 - 2nv_2^2) \end{bmatrix} \tag{12.2}$$

where $n = E_1/E_2$, $m = G_2/E_2$, which are named anisotropy parameters.

Hence, it is transformed into the x–y global coordinates as flows,

$$\{\sigma\} = [D]\{\varepsilon\} \tag{12.3}$$

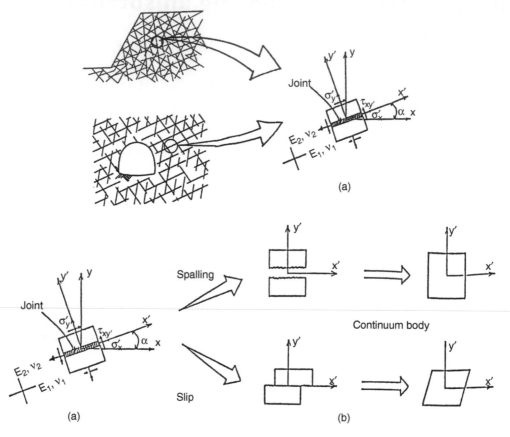

Figure 12.1 Modelling for discontinuous deformation in continuum mechanics.

where

$$[D] = [T][D'][T]^T \tag{12.4}$$

$[T]$ is a transformation matrix expressed as:

$$[T] = \begin{bmatrix} \cos^2\alpha & \sin^2\alpha & -2\sin\alpha\cos\alpha \\ \sin^2\alpha & \cos^2\alpha & 2\sin\alpha\cos\alpha \\ \sin\alpha\cos\alpha & -\sin\alpha\cos\alpha & \cos^2\alpha - \sin^2\alpha \end{bmatrix} \tag{12.5}$$

α is the angle between x'- and x-coordinate system.

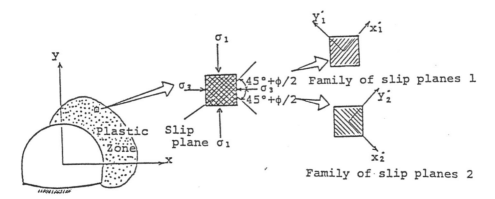

Figure 12.2 Families of slip planes in plastic zone.

12.3 DIFFERENT MODES OF DEFORMATION

It should be noted that Equation (12.3) can represent all three modes of deformation by changing the material constants, particularly the parameters n and m, which represent anisotropy of the materials.

12.3.1 Spalling of joints

The spalling of joints shown in Figure 12.1(b) can be represented by increasing the parameter n, i.e. by reducing the value of E_2 against E_1. Poisson's ratio v_2 is taken to be zero, because spalling in the direction of the y'-axis makes no movement in the direction of the x'-axis. In this case, the other parameter is taken as $m = 1/2(1 + v_1)$.

12.3.2 Sliding along joints

When sliding occurs along the joints parallel to the x'-axis, the parameter m can be reduced to a small value, i.e. $m < (1/2(1 + v))$, while $n = 1.0$ and $v_1 = v_2$ are assumed.

12.3.3 Plastic flow

Materials under a plastic state tend to slide along the two families of potential sliding planes with an angle of $\pm(45° + \varphi/2)$ from the maximum principal stress direction, as shown in Figure 12.2. φ denotes the internal friction angle.

Let us take two different coordinate systems for consideration of the mechanical behaviour of the two conjugate slip planes, as shown in Figure 12.2. The stress-strain relationship for each family of slip planes is given in the same form as Equations (12.1) and (12.3) for the local and global coordinate systems, respectively. The total strain is assumed to be expressed as:

$$\{\varepsilon\} = \frac{1}{2}\{\{\varepsilon_1\} + \{\varepsilon_2\}\} \tag{12.6}$$

where $\{\varepsilon_1\}$ and $\{\varepsilon_2\}$ are strains due to the families of two slip planes, respectively. Considering Equation (12.3), Equation (12.6) becomes:

$$\{\varepsilon\} = \frac{1}{2}[[D_1]^{-1} + [D_2]^{-1}]\{\sigma\} \tag{12.7}$$

This is the proposed stress-strain relationship for representing the plastic behaviour of materials. Equation (12.7) is expressed in the following common form:

$$\{\sigma\} = [D]\{\varepsilon\} \tag{12.8}$$

where

$$[D] = \left[\frac{1}{2}[[D_1]^{-1} + [D_2]^{-1}]\right]^{-1} \tag{12.9}$$

12.4   COMPUTER SIMULATIONS

In order to verify the validity of the constitutive equation for analysing tunnelling problems, computer simulations have been performed.

12.4.1   Spalling of joints

A tunnel under consideration here is shown in Figure 12.3. A main discontinuity plane is located 4 m above the tunnel crown. A joint element (Goodman et al., 1968) is used to represent the mechanical behaviour of the joints. In the computer simulation, a displacement distribution is obtained by an ordinary finite element analysis, where input data shown in Table 12.1 are used. The data for joints shown in the table are used only for the ordinary finite element analysis. The displacements along the discontinuous plane are shown in Figure 12.4. They clearly show the spalling of the joint due to tunnel excavations. The calculated displacements along the reference lines shown in Figure 12.3 are considered "measured" displacements. Back analysis for determining

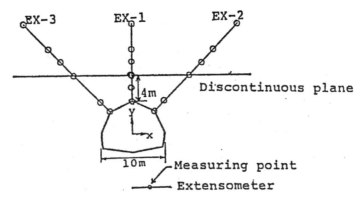

Figure 12.3 Tunnel excavated near a discontinuous plane and location of measuring points.

initial stress and mechanical properties from the "measured" displacements is carried out to minimise the error function given in Equation (2.3).

In the back analysis, two different constitutive equations are used, i.e., Equation (12.3) considering the anisotropy parameter n, and one for isotropic elastic material ($n = 1.0$).

The back analysis results are shown in Figure 12.5. It is obvious that the displacements calculated by considering the anisotropy parameters and the "measured" displacements coincide, while of course the isotropic elastic constitutive equation cannot simulate the spalling of joints.

Table 12.1 Input data and back analysis results in the case of spalling of joint.

| | FE analysis with joints | Back analysis | |
		Anisotropic	Isotropic
σ_x (MPa)	−2.45	−2.52	−0.75
σ_y (MPa)	−4.9	−4.9	−4.9
τ_{xy} (MPa)	0.0	0.0	0.0
E (MPa)	980	1034	670
Poisson's ratio v	0.3	0.3 (assumed)	
n	−	10.0	1.0

Data for joints

Wall rock compressive strength (MPa)	−6.86
Ratio of tensile to compressive strength	0.1
Shear stiffness (MPa)	377
Ratio of residual to peak shear strength	0.33
Maximum normal closure (cm)	10.0
Seating load (MPa)	−4.9
Friction angle of a smooth joint (degree)	30.0
Dilatancy angle (degree)	0.0

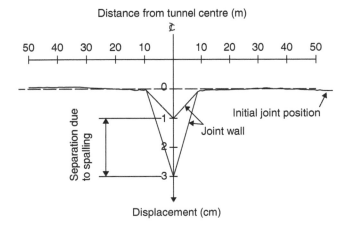

Figure 12.4 Spalling of joint.

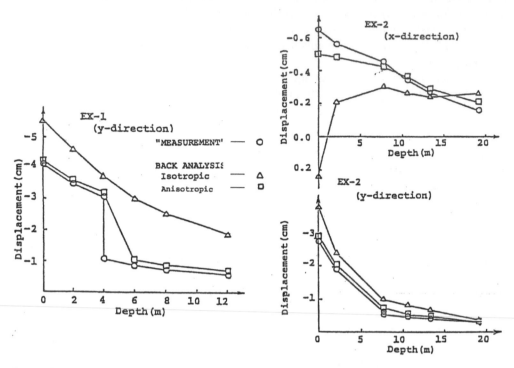

Figure 12.5 Comparison between measured and back-calculated displacements (Sakurai & Ine, 1986).

12.4.2 Plastic flow

It is assumed that a horseshoe-shaped tunnel, 5 m in radius, is bored in an elasto-plastic medium (see Figure 12.6). Elasto-plastic finite element analysis is first conducted by assuming the Drucker-Prager yield function together with a von Mises type of plastic potential function. The initial stress and material properties employed here are listed in Table 12.3. The calculated displacements along three reference lines, i.e. Ex-1, Ex-2 and Ex-3 (see Figure 12.6) are taken as "measured" displacements, which are used as input data for back analyses.

The plastic zone appearing around the tunnel is then back-calculated by the method proposed by Sakurai et al. (1985), and the result is compared with the "real" plastic zone as shown in Figure 12.6. The "real" plastic zone is one obtained by the ordinary elasto-plastic finite element analysis.

Once the plastic zone is back-calculated, the proposed back analysis is conducted to determine the initial stress and material constants including the anisotropic parameter m. It is noted in this analysis that the reduced parameter m is only valid in the plastic zone. The value $m = 1/2(1 + v)$ is taken in the region outside of the plastic zone. The anisotropic parameter m is described in detail in Section 14.3.

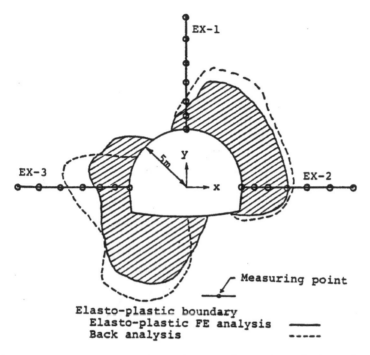

Figure 12.6 Horseshoe-shaped tunnel excavated in an elasto-plastic medium.

Table 12.2 Input data and back analysis results in the case of plastic flow.

| | Elasto-plastic FE analysis | Back analysis | |
		Proposed method	Isotropic
σ_x (MPa)	−2.45	−2.39	−2.78
σ_y (MPa)	−3.43	−3.43	−3.43
τ_{xy} (MPa)	0.85	1.10	0.98
E (MPa)	196	192	123
Poisson's ratio ν	0.3	0.3 (assumed)	
Cohesive strength C (MPa)	0.49	–	–
Internal friction angle φ (degree)	30.0	30.0	–
m	–	0.0962	$1/2\,(1+\nu) = 0.3846$
Critical strain ε_0 (%)	–	0.866	–

The back-calculated displacements along the three reference lines are shown in Figure 12.7 together with the "measured" displacements. In this figure the back analysis results obtained on the assumption of an isotropic elastic material are also shown for reference. It is obvious that the displacements back-calculated by the proposed method and the "measured" displacements coincide well with each other.

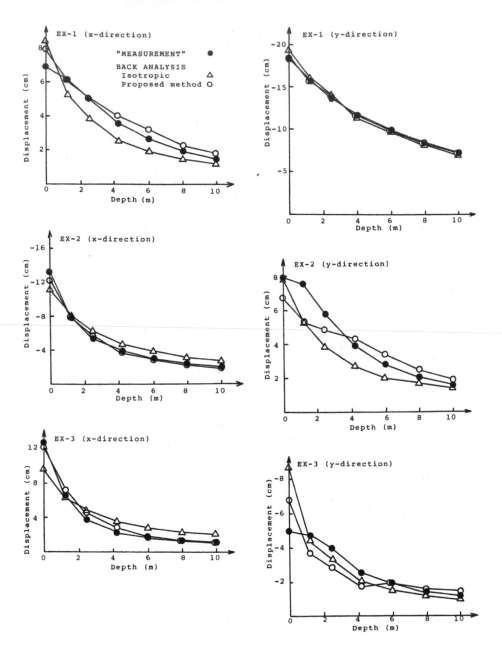

Figure 12.7 Comparison between measured and back-calculated displacements (Sakurai & Ine, 1986).

12.5 CASE STUDY (UNDERGROUND HYDROPOWER PLANT)

A large underground cavern for a hydroelectric power plant was constructed in a rock formation that consists mainly of tuff breccias and andesite. Several small shear

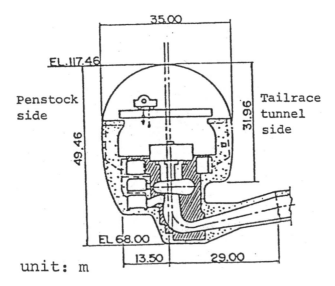

Figure 12.8 Cross section of the cavern.

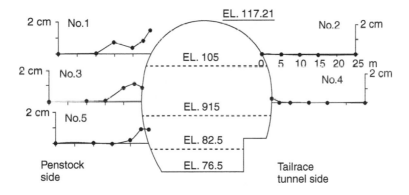

Figure 12.9 Location of multi-rod extensometers and measured displacements.

zones exist at the site of the powerhouse. Careful observations and field measurements were carried out during excavation for monitoring the stability of the cavern, and for verifying the adequacy of the design and the construction method (Sakurai & Tanigawa, 1989).

A cross section of the cavern is shown in Figure 12.8. The length of the cavern is 253 m. Displacement measurements were conducted during the excavation by use of multi-rod extensometers installed from the inner surface of the cavern. One of the locations where extensometers were installed is shown in Figure 12.9. The measurement results are also given in the same figure.

The finite element mesh used in the back analysis is shown in Figure 12.10. The loosened zones around the cavern are modelled as shown in Figure 12.11, where the

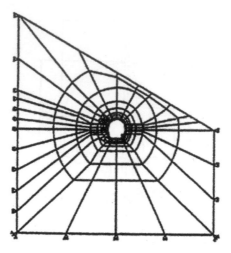

Figure 12.10 Finite element mesh.

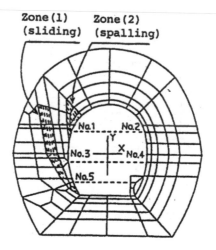

Figure 12.11 Sliding and spalling zones around the cavern.

sliding along potential slip planes and spalling of joints are assumed considering the results of geological explorations as shown in Figure 12.12. Considering the sliding and spalling of joints, the anisotropy parameters m and n in the loosened zone are back-calculated from the measured displacements to minimise the error function given in Equation (2.3), resulting in $m = 0.038$ and $n = 20$ being obtained (Sakurai & Tanigawa, 1989; Sakurai, 1991).

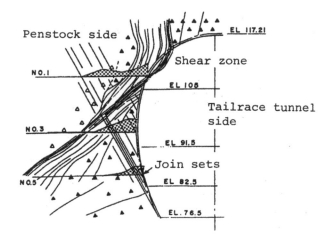

Figure 12.12 Geological conditions around the cavern.

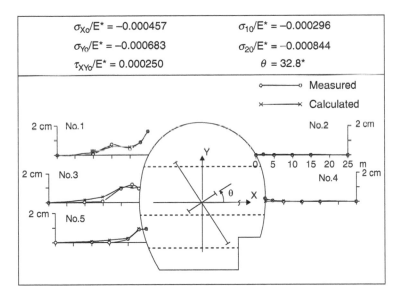

$\sigma_{X_0}/E^* = -0.000457$

$\sigma_{Y_0}/E^* = -0.000683$

$\tau_{XY_0}/E^* = 0.000250$

$\sigma_{1_0}/E^* = -0.000296$

$\sigma_{2_0}/E^* = -0.000844$

$\theta = 32.8^*$

Figure 12.13 Results of back analysis (anisotropic model).

In the back analyses, the normalised initial stress is also back-calculated from the measured displacements, which are used together with the back-calculated anisotropy parameters as input data for the ordinary (forward) finite element analyses, resulting in the displacements around the cavern being calculated, as shown in Figure 12.13. It shows a good agreement between the calculated and measured displacements. In the figure, the normalised initial stresses are listed.

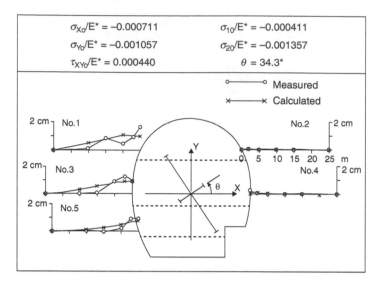

Figure 12.14 Results of back analysis (isotropic elastic model).

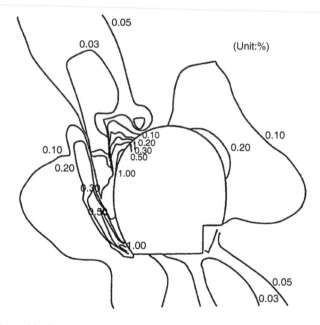

Figure 12.15 Maximum shear strain distribution (anisotropic model).

As a reference, the results of the back analysis carried out by assuming the isotropic linear elastic materials are shown in Figure 12.14. The maximum shear strain distribution in the case of using the anisotropy parameters is shown in Figure 12.15.

Chapter 13

Laboratory experiments

13.1 ABSOLUTE TRIAXIAL TESTS (TRUE TRIAXIAL TESTS)

In order to investigate the failure mechanism of rocks under a three-dimensional compressive stress state, the Absolute Triaxial Testing (ATT) machine was developed, and the absolute triaxial tests (more recently, this type of test is called a true triaxial test) were conducted by using a cubic specimen (Sakurai, 1966; Serata et al., 1968; Adachi et al., 1969). The special feature of the tests is the ability to control all three principal stresses independently, enabling the effect of the intermediate principal stress to be investigated. Cubic specimens were chosen as the standard laboratory specimens. To reduce the friction between the specimen and the loading plates as much as possible, the specimen was completely covered with friction reducers, which were made of several layers of Teflon sheet alternated with graphite grease. As rock specimens, rock salt was used because it is an homogeneous and isotropic material, so the failure mechanism of the rocks might be more easily investigated. A testing setup for a cubic specimen of rock salt is shown in Figure 13.1.

Cubic specimens (12.7 × 12.7 × 12.7 cm) of rock salt were tested under a triaxial compressive state. Figure 13.2 shows the specimens before and after testing. On the left of the figure is an intact specimen before testing; in the centre is a specimen tested under a triaxial stress state where the vertical stress was the maximum principal stress, while the two horizontal stresses, that is, the intermediate and the minimum principal stresses, were identical to each other; the specimen shown on the right is one tested under a uniaxial compressive stress state. Comparing a specimen tested under a triaxial compressive state (centre) with an intact specimen (left), it is obvious that the tested specimen is still solid, while the specimen tested under a uniaxial stress state (right) shows complete failure.

The test results demonstrate that no failure planes exist in the triaxial tests even though a large displacement occurs, while the specimen tested under a uniaxial compressive state fails due to separation rupture at the grain boundary. One of the experimental results of the absolute triaxial tests is shown in Figure 13.3, where the intermediate principal stress is approximately in the middle of the maximum and minimum principal stresses, while the maximum principal stress is a vertical stress. The shear failure occurs in a plane parallel to the direction of the intermediate principal stress, as seen in Figure 13.3, which indicates that the failure plane occurs at an angle

Figure 13.1 Setup of a rock salt specimen under triaxial loading conditions (Adachi et al., 1969).

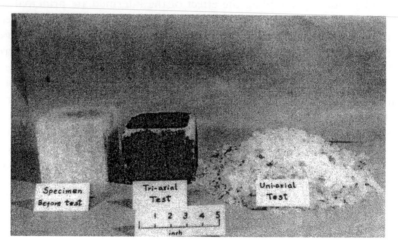

Figure 13.2 Specimens of rock salt: (a) before test; (b) after triaxial test; (c) after uniaxial test (Sakurai, 1966).

approximately $45° + \varphi/2$ (φ: internal friction angle) anticlockwise from the minimum principal stress direction (the horizontal in this figure).

The stresses at failure of the specimen are shown in the Mohr's stress circles (see Figure 13.4), which demonstrate that the Mohr–Coulomb failure criterion may hold for rock salt. The triaxial compressive tests where the three principal stresses were different from one another revealed that the intermediate principal stress has little effect on the failure of the material. This means that the failure criterion can be expressed in a two-dimensional way, in terms of only the maximum and minimum principal stresses.

Figure 13.3 Failure plane appearing on the surface perpendicular to the intermediate principal stress direction (Sakurai, 1966).

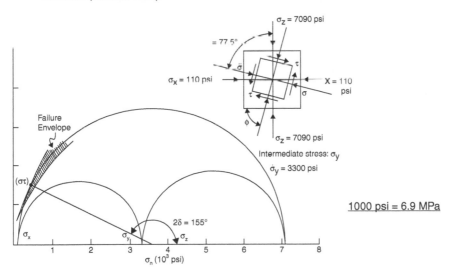

Figure 13.4 Mohr's failure envelope for determining the angle of the slip plane.

13.2 CONVENTIONAL TRIAXIAL COMPRESSION TESTS

In conventional triaxial tests, a cylindrical specimen is generally used to investigate the mechanical characteristics of geomaterials. However, there exists a shortcoming in the use of cylindrical specimens in that the intermediate principal stress is

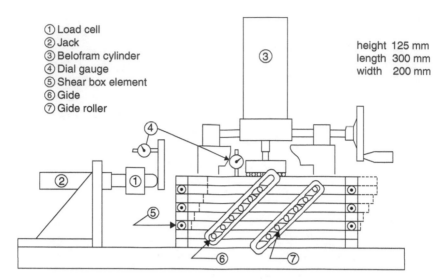

① Load cell
② Jack
③ Belofram cylinder
④ Dial gauge
⑤ Shear box element
⑥ Gide
⑦ Gide roller

height 125 mm
length 300 mm
width   200 mm

Figure 13.5 Simple shear-testing machine.

always identical to the minimum principal stress. Thus, failure planes do not appear in the specimen because failure planes are not uniquely determined, as discussed in Section 13.1.

Moreover, stress distributions in cylindrical specimens become very complex after the vertical stress reaches the yielding point of the materials. This problem is demonstrated by a numerical simulation in Chapter 15.

13.3 SIMPLE SHEAR TESTS

In order to investigate the non-linear mechanical behaviour of geomaterials under a shear stress state, a simple shear-testing machine has been developed, as shown in Figure 13.5. Using the testing machine, simple shear tests were carried out in the laboratory on sand in an unsaturated condition, and its mechanical characteristics have been carefully investigated. One of the results is shown in Figure 13.6, which indicates the relationship between shear stress and shear strain.

In the simple shear tests, a vertical force N was applied by loading and unloading at a certain level of shear strain that was kept constant during the vertical loading and unloading tests. The vertical loading and unloading were carried out several times at different levels of shear strain. Figures 13.7, 13.8 and 13.9 show the relationship between normal stress and normal strain in the vertical direction while the shear strain was kept constant at a given level, that is, $\gamma = 1.6\%$, $\gamma = 6.4\%$ and $\gamma = 16\%$, respectively. It can be seen from the figures that the normal stress-strain relationships in the vertical direction are almost identical for the three different shear strain levels. This means that the normal stress-strain relationship is independent of shear strains, so that Young's modulus is always constant no matter how large the shear strains applied, while it is obvious from Figure 13.6 that the shear modulus, defined as the inclination

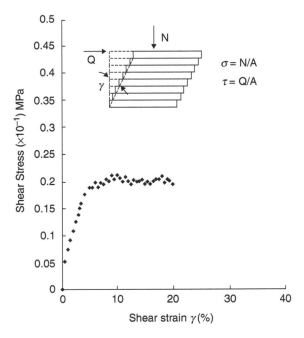

Figure 13.6 Shear stress and shear strain relationship as determined by a simple shear test (Sakurai & Akayuli, 1998).

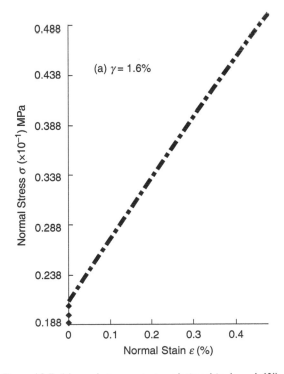

Figure 13.7 Normal stress–strain relationship ($\gamma = 1.6\%$).

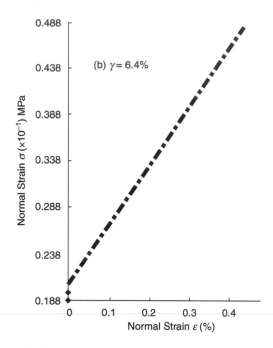

Figure 13.8 Normal stress–strain relationship ($\gamma = 6.4\%$).

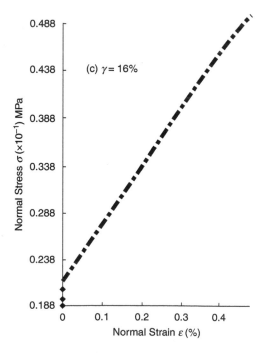

Figure 13.9 Normal stress–strain relationship ($\gamma = 16\%$).

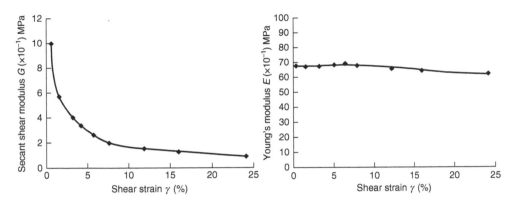

Figure 13.10 (a) Shear modulus vs. shear strain; (b) Young's modulus vs. shear strain.

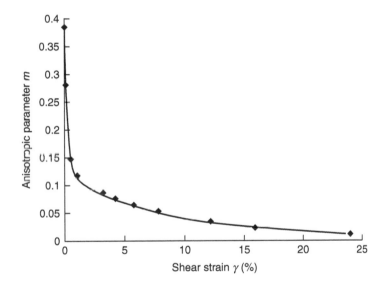

Figure 13.11 The ratio of shear modulus to Young's modulus (anisotropic parameter m, G/E) (Sakurai & Akayuli, 1998).

of the curve, decreases with increasing shear strain. Note: the secant shear modulus is adopted here.

The shear modulus, G, and Young's modulus, E, are plotted in relation to the shear strain as shown in Figures 13.10a and 13.10b, respectively. It is obvious from these figures that the shear modulus decreases with increasing shear strain, while Young's modulus is always constant no matter how large the shear strains applied. In other words, the non-linear behaviour of the materials is mainly caused by the reduction of shear rigidity (shear modulus G) along mobilised sliding planes, while Young's modulus, E, never varies, even after the shear stress exceeds the yielding point

(Sakurai & Akayuli, 1998). This must be an obvious result in the fundamental theory of solid mechanics, where Young's modulus and Poisson's ratio should always be constants because they are elastic material constants, while the non-linearity of the mechanical behaviours of materials is caused by the damage of the materials, resulting in the shear modulus decreasing as the shear strain increases.

In consideration of the results shown in Figures 13.10a and 13.10b, Sakurai (1987) proposed an "anisotropic parameter", which is defined as the ratio of the shear modulus to Young's modulus, that is, G/E, which is also plotted in relation to shear strain, as shown in Figure 13.11.

It should be noted that some researchers make the value of Young's modulus decrease with increase of stress levels after exceeding the yielding point. However, this is not theoretically correct, because Young's modulus is an elastic parameter and so should be constant until the failure of the material. This is clearly demonstrated in Figures 13.10(a) and (b), indicating that a non-linear relationship exists between shear stress and shear strain, while Young's modulus is always constant until a failure occurs in the material.

Chapter 14

Constitutive equations for use in back analyses

14.1 FUNDAMENTAL THEORY OF CONSTITUTIVE EQUATIONS FOR GEOMATERIALS

In numerical analyses, the constitutive equations for materials are extremely important because the accuracy and the reliability of calculation results depend entirely on what constitutive equations are used. In general, the constitutive equations for materials are developed on the basis of the theory of solid mechanics, particularly the theory of plasticity. However, it should be noted that the theory of plasticity was originally developed for metals, so that the associated flow rule holds in accordance with Drucker's definition for stable materials (Drucker 1951). In other words, both yield functions and plastic potential functions for metals are identical to each other. On the other hand, for frictional materials like rocks and soils, the results of laboratory experiments indicate that the plastic potential function differs from the yield function. Therefore, if the normality rule holds even for the frictional materials, the flow rule becomes a non-associated flow rule.

The non-associated flow rule requires both a plastic potential function and a yield function, which contain various mechanical parameters. In forward analyses, all these mechanical parameters can be evaluated prior to the design of the structures by performing laboratory and field tests, although it is not an easy task to determine all of them, because of the complexity of the geological and geomechanical characteristics of materials.

In back analyses, it is hard to adopt a complex constitutive equation, because the number of its mechanical parameters will be too large to back-calculate all of them from field measurement data, such as measured displacements. To overcome this difficulty, constitutive equations must be simple enough for all of the mechanical parameters to be determined from measured displacements. It should be noted that the constitutive equations used for back analyses do not necessarily follow the conventional theory of plasticity, but they could be derived on the basis of a completely different idea, only applied for back analyses.

14.2 FAILURE CRITERIA

14.2.1 Mohr–Coulomb failure criterion

The Mohr–Coulomb failure criterion is commonly used in the design of rock structures for assessing their stability. As already described, the intermediate principal stress has little effect on the failure of materials, so that the Mohr–Coulomb criterion can be expressed in a two-dimensional way in terms of only the maximum and minimum principal stresses.

Therefore, it is expressed in the following equation:

$$\tau = c + \sigma \tan \varphi \tag{14.1}$$

$$\text{where} \quad \tau = \frac{\sigma_1 - \sigma_3}{2} \cos \varphi \tag{14.2}$$

$$\sigma = \frac{\sigma_1 + \sigma_3}{2} - \frac{\sigma_1 - \sigma_3}{2} \sin \varphi \tag{14.3}$$

where σ_1 and σ_3 are the maximum and minimum principal stresses, respectively.

The Mohr–Coulomb criterion is expressed in the principal stress components as follows:

$$\sigma_1 - \sigma_3 = 2c \cos \varphi + (\sigma_1 + \sigma_3) \sin \varphi \tag{14.4}$$

14.2.2 Von Mises yield criterion

For ductile materials such as steel, the von Mises yield criterion is commonly used, in which the intermediate principal stress is taken into account:

$$\tau_{oct} = k \tag{14.5}$$

where τ_{oct} is the octahedral shear stress, defined as follows:

$$\tau_{oct} = \frac{1}{3} \sqrt{(\sigma_1 - \sigma_2)^2 + (\sigma_2 - \sigma_3)^2 + (\sigma_3 - \sigma_1)^2} \tag{14.6}$$

and k is a material constant.

14.2.3 Nadai's failure criterion and Drucker–Prager failure criterion

In order to extend the idea of von Mises yield criterion to frictional materials, Nadai (1950) proposed a failure criterion for rocks that takes into account the intermediate principal stress on the basis of the von Mises criterion, together with the Mohr–Coulomb criterion, as:

$$\tau_{oct} = F(\sigma_m) \tag{14.7}$$

where σ_m is the octahedral normal stress, expressed as:

$$\sigma_m = \frac{\sigma_1 + \sigma_2 + \sigma_3}{3} \tag{14.8}$$

The Drucker–Prager failure criterion was established as a generalisation of the Mohr–Coulomb criterion for soils (Drucker & Prager, 1952). It can be expressed as:

$$\sqrt{J_2} = \lambda I_1 + \kappa \tag{14.9}$$

where λ and κ are material constants, I_1 is the first invariant of the stress tensor and J_2 is the second invariant of the stress tensor deviator, which are defined as follows:

$$I_1 = \sigma_1 + \sigma_2 + \sigma_3 \tag{14.10}$$

$$J_2 = \frac{1}{6}[(\sigma_1 - \sigma_2)^2 + (\sigma_1 - \sigma_3)^2 + (\sigma_3 - \sigma_1)^2] \tag{14.11}$$

where σ_1, σ_2 and σ_3 are principal stresses.

The Drucker–Prager failure criterion, when expressed in terms of octahedral shear stress, τ_{oct}, and the mean stress, σ_m, is shown in the following form:

$$\tau_{oct} = \sqrt{\frac{2}{3}}(3\lambda\sigma_m + \kappa) \tag{14.12}$$

As seen in Equation (14.12), the Drucker–Prager failure criterion is based on the assumption that the octahedral shear stress at failure depends linearly on the mean stress through material constants. The Drucker–Prager failure criterion can be considered as a particular case of Nadai's criterion.

It should be noted that both Nadai's criterion and the Drucker–Prager criterion are established by using the octahedral shear stress acting on the octahedral planes. Therefore, these failure criteria cannot simulate the single sliding plane that appeared in the rock salts tested under a triaxial compressive stress state, as shown in Figure 13.3, because a sliding plane cannot be identified by the octahedral shear stress that acts on the octahedral planes.

Nevertheless, the failure criteria seem to be used as the failure criteria for rocks under a highly compressed stress state, that is, at a great depth. However, in underground structures such as tunnels, caverns and shafts, at least one of the principal stresses perpendicular to excavation surfaces must be small even though the structures are situated at great depth.

The Drucker–Prager failure criterion is often used for elasto-plastic numerical analyses in the design of tunnels, large caverns, vertical shafts, etc. The criterion is also formulated in terms of the octahedral shear stress acting on the octahedral planes, with the result that, as discussed in relation to Nadai's criterion, a single sliding plane cannot be simulated. In tunnels, it should be noted that at least one of the principal stresses near the inner surface of the tunnel must be small even though it is sited at a great depth, so that the sliding planes may occur. However, the sliding planes cannot be simulated by numerical analyses using the Drucker–Prager failure criterion.

14.3 ANISOTROPIC PARAMETER AND ANISOTROPIC DAMAGE PARAMETER

14.3.1 Anisotropic parameter

Based on experimental results, the anisotropic parameter, m, can be expressed as a function of the maximum shear strain, as shown in Figure 14.1.

As seen in Figure 14.1, the anisotropic parameter, m, is a simple monotonically decreasing function with respect to shear strain, and it can be expressed as a function of the maximum shear strain as follows:

$$m = f(\gamma) \tag{14.13}$$

where $f(\gamma)$ is expressed as follows:

$$f(\gamma) = G/E = 1/2(1+v) \quad \text{for } \gamma \leq \gamma_L \tag{14.14}$$

$$f(\gamma) = \frac{1}{2(1+v)} \exp\{-\alpha(\gamma - \gamma_L)\} \quad \text{for } \gamma > \gamma_L \tag{14.15}$$

where G: shear modulus; E: Young's modulus; v: Poisson's ratio; γ: maximum shear strain along mobilised sliding planes; γ_L: maximum shear strain at the elastic limit; α: a material constant.

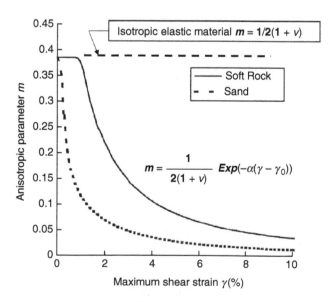

Figure 14.1 Anisotropic parameter m (= G/E) for rock and soil expressed in relation to the maximum shear strain (Sakurai et al., 2009).

14.3.2 Anisotropic damage parameter

In order to give the anisotropic parameter a more precise physical meaning, the anisotropic damage parameter, d, is defined (Sakurai & Akayuli, 1998):

$$d = \frac{1}{2(1+v)} - m \tag{14.16}$$

It is obvious from Equation (14.16) that the anisotropic damage parameter becomes $d = 0$ in an elastic state, where the anisotropic parameter is $m = 1/2(1+v)$. In other words, no damage occurs until the materials reach their yielding point. When the materials reach their yielding point, the anisotropic damage parameter starts to increase along mobilised slip planes, while the elastic parameters, such as Young's modulus and Poisson's ratio, always remain constant. Thus, the anisotropic damage parameter, d, implies the damage of the materials due to strain, that is, strain-induced damage of materials.

The anisotropic damage parameter, d, is expressed as a function of the maximum shear strain in the same fashion as the anisotropic parameter, as previously shown in Equations (14.13) and (14.14).

$$d = g(\gamma) \tag{14.17}$$

where

$$\left.\begin{aligned}
g(\gamma) &= 0 & \text{for } \gamma \leq \gamma_L \\
g(\gamma) &= \left(\frac{1}{2(1+v)} - \beta\right)[1 - Exp\{-\alpha(\gamma - \gamma_0)\}] & \text{for } \gamma > \gamma_L
\end{aligned}\right\} \tag{14.18}$$

where α and β are material constants.

As seen in Equation (14.18), the function $g(\gamma)$ is a monotonically increasing function with the maximum shear strain γ, with the result that the anisotropic damage parameter, d, increases monotonically with an increase in the maximum shear strain.

14.4 PROPOSED CONSTITUTIVE EQUATION FOR GEOMATERIALS

14.4.1 Constitutive equation

On the basis of the results of laboratory experiments on rock salt (Sakurai, 1966), it was found that the intermediate principal stress has little effect on a failure, as described in Section 13.1. Therefore, it is necessary to consider only the maximum and minimum principal stresses. As a result, the intermediate principal stress direction is taken as the z-axis, which is perpendicular to the x-y coordinate system, and a two-dimensional plane strain analysis can be performed. The local x' coordinates are then taken along the potential slip plane, along which the anisotropic parameter, m, is taken into account, which can be back-calculated from measured displacements, so as to minimise the discrepancy between the measured and the back-calculated displacements.

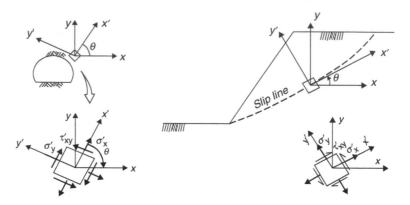

Figure 14.2 Local and global coordinates.

Considering the anisotropic parameter, Sakurai (1987) proposed a constitutive equation for a two-dimensional plane strain state, as given in Equation (14.19). The constitutive equation is valid in the local coordinate system x′–y′, where the x′-axis is parallel to a potential slip plane, as shown in Figure 14.2.

$$\left\{ \begin{array}{c} \sigma_{x'} \\ \sigma_{y'} \\ \tau_{x'y'} \end{array} \right\} = [D'] \left\{ \begin{array}{c} \varepsilon_{x'} \\ \varepsilon_{y'} \\ \gamma_{x'y'} \end{array} \right\} \tag{14.19}$$

where

$$[D'] = \frac{E}{1 - v - 2v^2} \begin{bmatrix} 1 - v & v & 0 \\ v & 1 - v & 0 \\ 0 & 0 & m(1 - v - 2v^2) \end{bmatrix}$$

where E: Young's modulus; v: Poisson's ratio; m: anisotropic parameter ($m = 1/2(1 + v) - d$); d: anisotropic damage parameter.

It is obvious that if anisotropic damage parameter $d = 0$, that is, no damage, then the material is an isotropic linear elastic material, while if $d \neq 0$ the mechanical behaviour of the material must be non-elastic. This means that the constitutive equation given in Equation (14.19) can simulate any geomaterials whose mechanical characteristics range from isotropic linear elastic materials to elasto-plastic non-linear materials. Therefore, if we use the proposed constitutive equation, there is no need to assume any mechanical models for back analyses. The result is that the back analysis results obtained by using the constitutive equation reveal whether the material is still in an elastic state, or is already in a non-elastic (plastic) state, depending on the value of the anisotropic damage parameter. This fact represents a great advantage for back analyses, because we do not have to worry about what types of mechanical model of the materials are adopted prior to the back analyses, with the result that a mechanical model should not be assumed, but should be determined by back analysis.

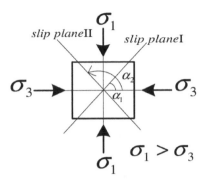

Figure 14.3 Two conjugate slip planes.

The constitutive equation given in Equation (14.19) can be transformed into the global coordinate system as the following equation:

$$\begin{Bmatrix} \sigma_x \\ \sigma_y \\ \tau_{x'y} \end{Bmatrix} = [D] \begin{Bmatrix} \varepsilon_x \\ \varepsilon_y \\ \gamma_{x'y} \end{Bmatrix}$$

(14.20)

where

$$[D] = [T][D'][T]^T$$

(14.21)

and where $[T]$ is a transformation matrix expressed as:

$$[T] = \begin{bmatrix} \cos^2\theta & \sin^2\theta & -2\sin\theta\cos\theta \\ \sin^2\theta & \cos^2\theta & 2\sin\theta\cos\theta \\ \sin\theta\cos\theta & -\sin\theta\cos\theta & \cos^2\theta - \sin^2\theta \end{bmatrix}$$

(14.22)

where the angle θ is expressed as the angle between the x- and x'-axes, as shown in Figure 14.2.

It should be noted that there exist, in general, two conjugate potential slip planes at a certain point within materials under a given stressed state. It is seen from Figure 14.3 that one is located at an angle of α_1, measured anticlockwise from the direction of minimum principal stress (compression is positive), and the other is at an angle of α_2. The angles of α_1 and α_2 are determined for geomaterials that obey the Mohr–Coulomb failure criterion as follows:

$$\alpha_1 = 45° + \frac{\varphi}{2}, \quad \alpha_2 = 135° - \frac{\varphi}{2}$$

(14.23)

where φ is the internal friction angle of the geomaterial.

Considering the slip planes I and II shown in Figure 14.3, the strain due to each slip plane can be expressed as:

$$\{\varepsilon_1\} = [D_1]^{-1}\{\sigma\} \tag{14.24}$$

$$\{\varepsilon_2\} = [D_2]^{-1}\{\sigma\} \tag{14.25}$$

The total strain can then be written as:

$$\{\varepsilon\} = \frac{1}{2}\{\{\varepsilon_1\} + \{\varepsilon_2\}\} \tag{14.26}$$

The stress–strain relationship representing the mechanical behaviour of materials in the global coordinate system is then expressed as:

$$\{\sigma\} = [D]\{\varepsilon\} \tag{14.27}$$

In the two-dimensional form, Equation (14.27) can be expressed as follows:

$$\begin{Bmatrix} \sigma_x \\ \sigma_y \\ \tau_{xy} \end{Bmatrix} = [D] \begin{Bmatrix} \varepsilon_x \\ \varepsilon_y \\ \gamma_{xy} \end{Bmatrix} \tag{14.28}$$

Combining Equations (14.24) through (14.27) provides the complete stress–strain relationship in the global coordinate system, as shown in Equation (14.28), where the [D] matrix is expressed in the following form:

$$[D] = \left[\frac{1}{2}([D_1]^{-1} + [D_2]^{-1}) \right]^{-1} \tag{14.29}$$

where $[D_1]$ and $[D_2]$ are the stiffness matrices for two conjugate potential slip planes I and II, respectively, which are calculated by Equation (14.19).

It should be noted that since the anisotropic parameter, m, is a function of the maximum shear strains, as shown in Figure 14.1, it should be back-calculated from measured displacements by an iterative computing procedure in which the discrepancy between the measured displacements and those derived from a numerical analysis is minimised, as shown in Equation (2.3). In the iterative procedure, both Young's modulus and Poisson's ratio should also be back-calculated from measured displacements.

Even though the constitutive equation given in Equation (14.19) is expressed in a two-dimensional plane strain state, it could be easily transformed to a three-dimensional state. However, it is not necessary to do so because, as already discussed in Section 13.1, the intermediate principal stress has little effect on a failure, with the result that only the maximum and minimum principal stresses need be considered in back analyses. As a result, the two-dimensional constitutive equation can be used in engineering practice.

Another problem in the use of Equation (14.19) may be the dilatancy effect, which is a typical mechanical characteristic of geomaterials such as soils and rocks. The dilatancy effect is the volume change of materials due to the application of shear stresses.

Looking at the $[D']$ matrix given in Equation (14.19), it seems that the dilatancy effect is disregarded. However, if necessary, the dilatancy effect can be taken into account in back analyses by considering the non-elastic volumetric strains, which are described in Section 9.2. The non-elastic volumetric strains can be back-calculated from the measured displacements through consideration of Equations (9.6) and (9.7). The approach to considering non-elastic strains in back analyses is described in Chapter 9.

14.4.2 Objectivity of constitutive equation

Objectivity must always be satisfied in any kind of constitutive equation. In other words, any constitutive equations should be tensor equations, with the result that they can be transformed to any other coordinate system with their physical meaning unchanged. The simplest example of a constitutive equation satisfying this objectivity is Hooke's law for an isotropic elastic material, where the only mechanical constants are Young's modulus and Poisson's ratio, which never change under the transformation of coordinates to any other coordinate system.

Regarding the constitutive equation that considers the anisotropic parameter shown in Equation (14.19), it has three mechanical constants, that is, Young's modulus, Poisson's ratio and the anisotropic parameter. In Hooke's law, the shear modulus, G, is not an independent mechanical parameter, but it is always related to Young's modulus and Poisson's ratio by the following form:

$$G = \frac{E}{2(1 + v)} \tag{14.30}$$

where E: Young's modulus; v: Poisson's ratio.

The $[D']$ matrix given in Equation (14.19) is implicitly defined such that the shear modulus, G, is no longer proportionally related to Young's modulus, that is, the relationship between Young's modulus, E, and shear modulus, G, is expressed in the following inequality expression:

$$G \le \frac{E}{2(1 + v)} \tag{14.31}$$

This means that the value of the anisotropic parameter is always smaller than $1/2(1 + v)$, as shown in the following inequality expression:

$$m \le \frac{1}{2(1 + v)} \tag{14.32}$$

In the constitutive equation shown in Equation (14.19), the mechanical parameters are Young's modulus, Poisson's ratio and the anisotropic parameter. Thus, the materials are no longer isotropic elastic ones, with the result that there is no guarantee of the objectivity of the constitutive equations being satisfied. Considering these situations, Sakurai et al. (1986) proved that the objectivity of the constitutive equation is approximately satisfied, so that it can be applied to any problem in engineering practice. It should be noted that the anisotropic parameter, m, could be reduced to $m = 0$, from an objectivity point of view.

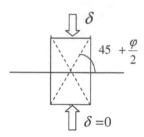

Figure 14.4 Uniaxial compressive test.

The constitutive equation that considers the three different deformational modes, that is, spalling, sliding and plastic flow, is described in Equation (12.1), where the anisotropy parameter $n = E_1/E_2$ is considered in addition to the anisotropic parameter, m. From the objectivity point of view, caution must be paid to the anisotropy parameter, n, which may have a particular limitation such that it cannot be larger than a certain value.

14.5   APPLICABILITY OF THE PROPOSED CONSTITUTIVE EQUATION

The applicability of the proposed constitutive equation is demonstrated by showing a numerical simulation, in which an elasto-plastic material is compressed under a uniaxial compressive stress state. The uniaxial compressive loading tests, so-called strain-controlled compression tests, are carried out in such a way that the vertical displacement, δ, is applied on the top of the specimen, while the bottom of the specimen is fixed in the vertical direction (see Figure 14.4). The numerical simulation is done in a two-dimensional plane strain condition (Sakurai & Shinji, 2005).

In order to determine the mechanical characteristics of materials, laboratory experiments must be performed. However, since the main purpose of these numerical simulations is to demonstrate the applicability of the anisotropic damage parameter and/or anisotropic parameter, the mechanical parameters of the geomaterials are assumed, as follows:

Young's modulus: $E = 1000\,\text{MPa}$
Poisson's ratio: $\nu = 0.3$
Critical shear strain: $\gamma_0 = 1.34\%$
Internal friction angle: $\varphi = 20°$

The stress–strain relationship of the geomaterials is also assumed, as shown in Figure 14.5. As seen in the figure, four different cases are considered, that is, strain hardening (case 1), perfectly plastic (case 2) and strain softening (cases 3 and 4). The stress–strain relationship is used as input data for the numerical simulation.

Taking the mechanical parameters of the geomaterials, together with the stress–strain relationship shown in Figure 14.5, as input data, a finite element analysis is performed to back-calculate the anisotropic damage parameter. In the calculation, the

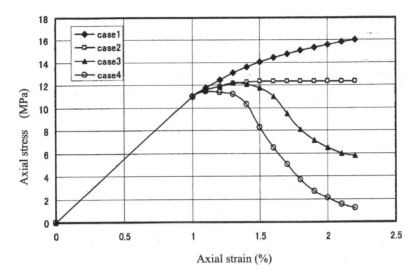

Figure 14.5 Stress–strain relationship.

vertical displacement corresponding to the elastic limit is easily calculated by a linear elastic model in which the anisotropic damage parameter is zero. After the elastic limit, as the vertical displacement continuously increases, some damage starts occurring in the materials. The anisotropic damage parameter can then be back-calculated by a direct approach, as described in Section 2.3.3, in such a way that a certain amount of the increment of the vertical displacements is applied, and the anisotropic damage parameter is back-calculated in an iterative process until the point of intersection of axial stress with axial strain lies on the line of stress–strain relationship, given in Figure 14.5. This calculation procedure is repeated for each case by increasing the vertical displacements step by step. The final results of the anisotropic damage parameter, back-calculated from the application of vertical displacements, are plotted in relation to the vertical strain, as shown in Figure 14.6. The anisotropic damage parameter shown in Figure 14.6 is the result of the back analysis of the numerical simulation (in Figure 14.6, the anisotropic parameter is also plotted on the vertical axis on the right-hand side).

In comparing Figure 14.5 with Figure 14.6, it is of interest that the anisotropic damage parameter can simulate the three different types of non-linear mechanical characteristic of geomaterials, such as strain-hardening, perfectly plastic and strain-softening behaviours, by a simple monotonic increasing function with only a slight change in the strain rate. Because the anisotropic damage parameter of geomaterials is expressed by a simple monotonic increasing function, the anisotropic damage parameter of the geomaterials can easily be back-calculated from measured displacements without assuming any non-linear mechanical model of the geomaterials.

This means that if the anisotropic (damage) parameter is used as input data for a forward analysis, it is not necessary to assume any type of non-linear mechanical model prior to back calculations, because the back-calculated anisotropic parameter

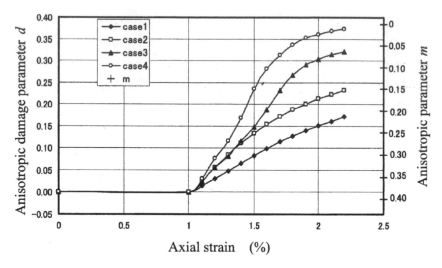

Figure 14.6 The anisotropic damage parameter and anisotropic parameter related to axial stress.

can identify the mechanical model of geomaterials implicitly from measured displacements (in this numerical simulation, the stress–strain relationship is provided as a measured value). In engineering practice, for example in a tunnel construction, the anisotropic (damage) parameter is back-calculated from measured displacements, and it is used as input data for forward analyses for determining the maximum shear strain distribution around the tunnel. The maximum shear strains are then compared with the critical shear strain of the geomaterials to assess the stability of the tunnel. In this back analysis, there is no need to assume the mechanical model of the geomaterials, but it is implicitly determined by the back-calculated anisotropic (damage) parameter.

The anisotropic parameter is also presented in Figure 14.6, together with the anisotropic damage parameter. However, the latter is much more useful in engineering practice, because it can easily be confirmed whether the geomaterials are still in an elastic state, or already in a plastic state, by the value of the anisotropic damage parameter. If the value is zero, then the geomaterials are still in an elastic state, while if its value starts increasing from zero, the geomaterials move into a plastic state. Therefore, it may be possible for the anisotropic damage parameter to be used to assess how much damage such as a non-linear mechanical behaviour starts occurring. In this regard, the anisotropic damage parameter seems to be much more useful than the anisotropic parameter in engineering practice.

However, the anisotropic parameters defined by the ratio of shear modulus to Young's modulus may make it easier for practising engineers to understand what the anisotropic parameter is from the physical point of view. As a reference, Figure 14.6 is shown in a different presentation in Figure 14.7, where the anisotropic damage parameters are plotted in relation to the maximum shear strain. This figure is the same as that shown in Figure 14.1, even though it is upside down.

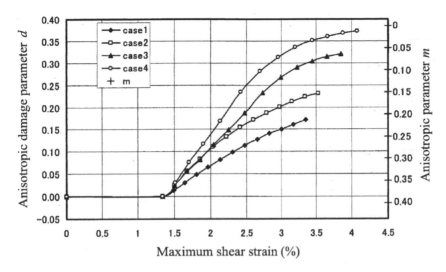

Figure 14.7 The parameters *d* and *m* vs. the maximum shear strain.

14.6   CONCLUSIONS ON THE RESULTS OF THE NUMERICAL SIMULATION

The numerical simulation demonstrates that the proposed constitutive equation given in Equation (14.19) can be highly applicable to back analysis of measured displacements during the construction of tunnels. The main purpose of back analyses is to verify the adequacy of the original design, and also to assess the stability of tunnels during their construction. For these purposes, the maximum shear strains around the tunnels must be determined by back analyses of measured displacements, and if the back-calculated maximum shear strains are still smaller than the critical shear strain, then the tunnels are stable; if not, the original design must be modified. In this assessment procedure, the stress–strain relationship is not needed. This means that it is not necessary to identify whether strain-hardening, perfectly plastic or strain-softening behaviour is occurring. It should be noted that the non-linearity of the mechanical characteristics of geomaterials is implicitly considered in back analysis, provided that the anisotropic (damage) parameter is used.

In the proposed back analyses, the anisotropic (damage) parameter is a powerful tool, which can easily simulate strain-hardening, perfectly plastic and strain-softening behaviours of the geomaterials without the assumption of any mechanical model. Furthermore, it is an advantage that the anisotropic (damage) parameters are expressed in a simple monotonic increasing function with respect to the maximum shear strain, with only a slightly change to the strain rate, as shown in Figure 14.7. This may result in the anisotropic (damage) parameter of the geomaterials being easily back-calculated from measured displacements, without assuming any non-linearity in the mechanical characteristics of the geomaterials.

In engineering practice for tunnels, the observational method is popularly used during construction. Field measurements, particularly displacement measurements, are commonly performed, and the measurement data must be properly interpreted, which is where back analysis becomes a powerful tool. However, it is extremely difficult to identify a mechanical model of geomaterials by a back analysis of measured displacements, whereas if the anisotropic (damage) parameter is introduced there is no need to assume any mechanical model, with the result that a mechanical model can be determined implicitly from measured displacements. Therefore, the maximum shear strain around the tunnel can be easily back-calculated from measured displacements without assuming any mechanical model, and the back-calculated maximum shear strain can be compared with the critical shear strain of the geomaterials to assess the stability of the tunnel.

14.7 FORWARD ANALYSIS VS. BACK ANALYSIS

It should be noted that non-linear mechanical behaviours of geomaterials, such as strain hardening, perfectly plastic and strain softening, are observed in triaxial compression tests in the laboratory, depending on the magnitude of the confining pressures. In forward analyses, any type of mechanical model can be established. However, it is extremely difficult to determine all of the mechanical parameters for the non-linear mechanical characteristics of geomaterials, even though *in situ* tests are performed.

In the conventional elasto-plastic numerical analyses carried out at the design stage of rock structures, a perfectly plastic model is usually adopted because of its simplicity in computational algorithms, while strain-hardening and strain-softening models are scarcely used. This may be due to the fact that it is extremely difficult to determine the mechanical models expressing both strain-hardening and strain-softening behaviours. In addition, many engineers have been using computer code that implements a perfectly plastic mechanical model for the design of rock structures. Various numerical analysis procedures have already been proposed for expressing the perfectly plastic behaviour of the materials, which are mostly based on the conventional theory of plasticity.

On the other hand, in back analyses it is extremely difficult to back-calculate all of the mechanical parameters from measured displacements, because the number of mechanical parameters contained in the mechanical model is usually too large to determine all of them by back analyses. Considering these difficulties for back analyses, it should be emphasised that a mechanical model must be simple enough to be able to determine all of the mechanical parameters by a limited number of measured displacements. To achieve this requirement, the constitutive equation given in Equation (14.19) must be useful because the number of mechanical parameters of the equation is only three, that is, the anisotropic (damage) parameter, Young's modulus and Poisson's ratio (being assumed), so that it is not difficult to back-calculate all of the parameters from measured displacements. In addition, the anisotropic (damage) parameter is a simple monotonic function with respect to the maximum shear strains, so that it is easily back-calculated from the limited number of displacement data measurements. Furthermore, the anisotropic (damage) parameters can easily simulate all three non-linear mechanical behaviours, that is, strain hardening, perfectly plastic and strain softening, without assuming any mechanical model.

One of the objectives of back analyses in engineering practice is to re-evaluate the mechanical parameters of rock masses used at the design stage. In a back analysis, however, the mechanical model (constitutive equation) of a rock mass is usually assumed to be the same as that used in the forward analysis performed at the design stage. For instance, if an isotropic linear elastic mechanical model is assumed, Young's modulus and Poisson's ratio are back-calculated. However, if an elasto-plastic model is assumed, then the mechanical parameters of cohesion and the internal friction angle are obtained in addition to Young's modulus and Poisson's ratio, even though the identical input data (measurement data) are used. It should be noted that back analyses can derive the correct answer for the mechanical parameters of geomaterials only if the assumed mechanical model is "true". However, it is almost impossible to assume a real mechanical model for geomaterials. This implies that the mechanical model should not be assumed, but should be back-calculated from measured displacements. Consequently, it is emphasised that the constitutive equations used at the design stage should not be used for back analyses in engineering practice.

Chapter 15

Cylindrical specimen for the determination of material properties

15.1 INTRODUCTION

Various constitutive equations for geomaterials have been proposed. The mechanical parameters of constitutive equations are determined by laboratory and *in situ* tests. This type of experiment is called an "element test", which requires a uniform and homogeneous distribution of stresses in a specimen during loading, even in a post-yield condition. A cylindrical specimen is commonly used for both uniaxial and triaxial tests. It may be thought, however, that slip planes start to mobilise in the specimen, with the result that the stress distribution in the specimen may become non-uniform. This means that care must be taken for the element tests, particularly when determining post-yield mechanical parameters. As an example of the applicability of the proposed constitutive equation, that is, Equation (14.19), the stress distribution in a cylindrical specimen is calculated by a forward analysis, and the adequacy of the use of cylindrical specimens is discussed in the context of an element test (Sakurai & Shinji, 2008).

15.2 CONSTITUTIVE EQUATION FOR CYLINDRICAL COORDINATE SYSTEMS

The constitutive equation given for a two-dimensional plane strain condition, that is, Equation (14.19), can be extended into the cylindrical coordinate system, as follows:

$$\{\sigma'\} = [D']\{\varepsilon'\} \tag{15.1}$$

where

$$\{\sigma'\} = \begin{Bmatrix} \sigma_r \\ \sigma_\theta \\ \sigma_z \\ \tau_{rz} \end{Bmatrix} \quad \text{and} \quad \{\varepsilon'\} = \begin{Bmatrix} \varepsilon_r \\ \varepsilon_\theta \\ \varepsilon_z \\ \gamma_{rz} \end{Bmatrix} : \tag{15.2}$$

$$[D'] = \frac{E}{(1+v)(1-v-2v^2)} \begin{bmatrix} 1-v^2 & v(1+v) & v(1+v) & 0 \\ & 1-v^2 & v(1+v) & 0 \\ & & 1-v^2 & 0 \\ sym & & & m(1+v)(1-v-2v^2) \end{bmatrix} \tag{15.3}$$

where E: Young's modulus; v: Poisson's ratio; m: anisotropic parameter ($m = 1/2(1 + v) - d$).

15.3 NUMERICAL SIMULATION

15.3.1 Introduction

In order to investigate the stress distribution of specimens tested under a uniaxial compressive state, numerical simulations are carried out. In the numerical simulations, three different shapes of specimens are tested, that is, rectangular solid specimens, cylindrical specimens and hollow cylindrical specimens. All the specimens are tested under a constant displacement control loading condition. The rectangular solid specimen is tested under a plane strain condition. The finite element method is used in the simulations, where the anisotropic parameter is used in expressing the post-yield mechanical behaviour of the specimens. The stress distributions in these three different types of specimens are calculated, and the results are discussed from the viewpoint of the uniformity of stress distributions in the specimens.

15.3.2 Stress distribution in differently shaped specimens

Three differently shaped specimens, namely, (1) rectangular solid, (2) cylindrical, (3) hollow cylindrical, are used for the numerical simulations. All of the specimens are compressed beyond the elastic limit of the materials under constant vertical displacements. The anisotropic parameter, m, is considered in the calculations. The coordinate systems for these three specimens are shown in Figure 15.1.

The sizes of the specimens used for the numerical simulations are:

1 Rectangular solid specimen: height $H = 10$ cm; width $W = 2.5$ cm.
2 Cylindrical specimen: height $H = 10$ cm; diameter $\varphi = 5$ cm.
3 Hollow cylindrical specimen: height $H = 10$ cm; diameter $\varphi = 5$ cm, with a hole of diameter $\varphi = 1$ cm located at its centre.

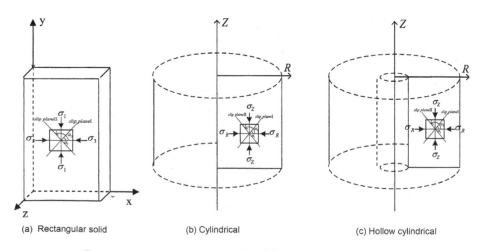

(a) Rectangular solid (b) Cylindrical (c) Hollow cylindrical

Figure 15.1 Coordinate systems for differently shaped specimens.

15.3.3 Principal stress distributions

The principal stress distributions for both the rectangular solid and the cylindrical specimens are shown in Figure 15.2, where Young's modulus $E = 1000$ MPa, and Poisson's ratio $v = 0.3$. In the numerical simulation the anisotropic parameter $m = 0.1$ is used. This means that the specimens have already been in a plastic state. It can be seen from the figure that the stress distribution for the rectangular solid specimen is uniform and homogeneous, while the stress distribution for the cylindrical specimen is no longer uniform in a plastic state.

15.3.4 Distribution of stress components along a given cross section

It is obvious that stresses in the three different types of specimens, that is, rectangular solid, cylindrical and hollow cylindrical, are uniformly distributed when they are compressed within the elastic limit. However, for both the cylindrical and hollow cylindrical specimens, stress distributions are no longer uniform when the specimens are compressed beyond the elastic limit, as seen in Figure 15.2, while for the rectangular solid specimen the stresses are uniformly distributed, even though the specimens are already in a plastic state. On the other hand, the stress components along the radial

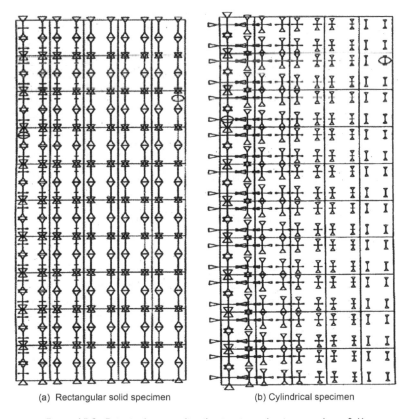

(a) Rectangular solid specimen   (b) Cylindrical specimen

Figure 15.2 Principal stress distribution in a plastic state ($m = 0.1$).

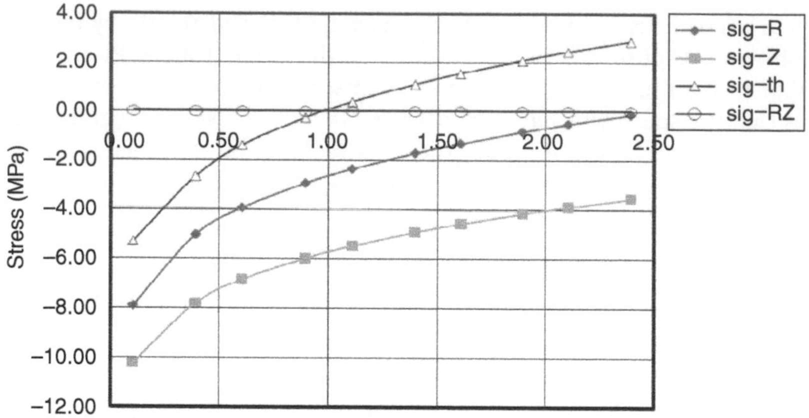

(a) Cylindrical specimen ($m = 0.1$)

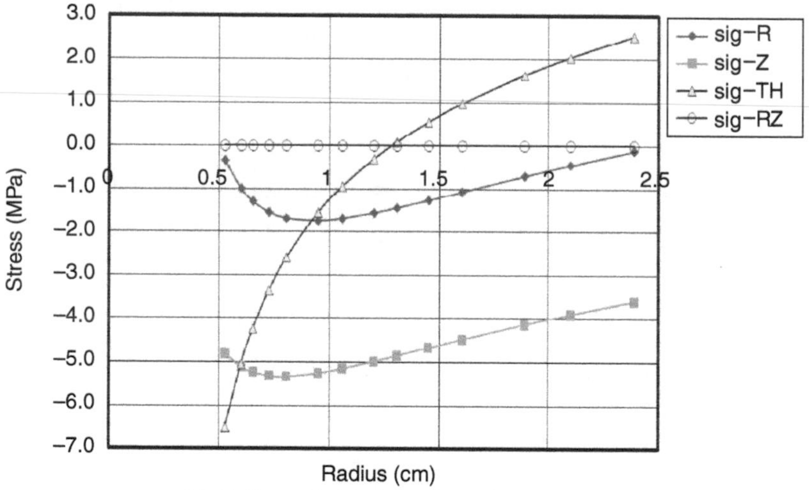

(b) Hollow cylindrical specimen ($m = 0.1$)

Figure 15.3 Distribution of the stress components ($\sigma_r, \sigma_\theta, \sigma_z, \tau_{rz}$).

direction for a given vertical cross section of both the cylindrical and the hollow cylindrical specimens are shown in Figure 15.3. These stress distributions are obtained by considering the anisotropic parameter $m = 0.1$, which means that the specimens have already been in a plastic state. It can be seen from the figure that the stresses are not uniformly distributed for either specimen.

15.3.5 Discussions/conclusions

It is obvious that the stress distributions of specimens compressed within the elastic limit are homogeneous and uniform no matter what shapes of specimens are used.

However, when the cylindrical and the hollow cylindrical specimens are compressed beyond the elastic limit, the stress distributions are no longer uniform or homogeneous, as shown in Figures 15.2 and 15.3. It is demonstrated that the stress distributions in a specimen depend entirely on the specimen's shape. On the other hand, it is of interest to know that the stresses in the rectangular solid specimen are always homogeneous and uniform even beyond the elastic limit, as shown in Figure 15.2a.

It is common to use a cylindrical specimen to determine the compressive strength and deformability, such as Young's modulus, of geomaterials. However, on the basis of the foregoing discussion, the use of cylindrical specimens in laboratory tests must be questionable. In contrast, if a rectangular solid specimen is tested under a plane strain condition, the stress distribution of the specimen is uniform and homogeneous even after the elastic limit. Thus, rectangular solid specimens are preferable in determining the post-yield mechanical parameters of constitutive equations for geomaterials (Sakurai & Shinji, 2008).

Considering the evidence brought to light by the numerical simulation results, the fact that a cylindrical specimen is commonly used in conventional laboratory tests is called into question when determining the mechanical parameters of geomaterials in a plastic state, whereas laboratory tests using cylindrical specimens may be meaningful as physical models for investigation of the stability of pillars in a room-and-pillar excavation method in underground mines.

Chapter 16

Applicability of anisotropic parameter for back analysis

16.1 PHYSICAL MODEL TESTS IN LABORATORY

In order to demonstrate the applicability of the proposed constitutive equation to back analyses for assessing the stability of tunnels from measured displacements, a physical model test was carried out. A model of two parallel tunnels at shallow depth was set up by piling up aluminium bars around the tunnels containing airbags at a pressure equal to the overburden pressure. The aluminium bars were 50 mm long with diameters of 3.0 mm and 1.6 mm, mixed in a ratio of 2:3 to simulate sand. Each of the tunnels were of diameter $D = 15$ cm and were separated from each other by a 0.5D-wide pillar with an overburden of 1D. Horizontal and vertical lines were drawn across the surface of the model's aluminium bars to form a 1 cm square grid. At the spring-line level of the tunnels, small load cells were inserted in both the pillar and the sides of the tunnel for the measurement of incremental stresses as the airbag pressure was reduced. A displacement measurement sensor (Keyence LB-1000 laser unit) was also mounted at the top of the setup to measure surface subsidence. Figures 16.1 and 16.2 show the laboratory setup of the two-tunnel system (Akayuli & Sakurai, 1998).

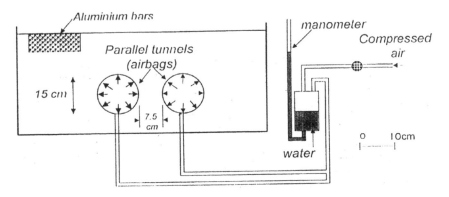

Figure 16.1 Physical model of two tunnels.

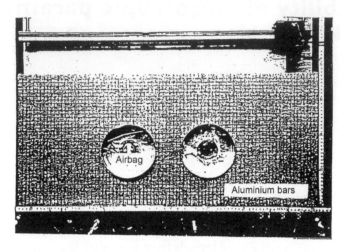

Figure 16.2 Laboratory model setup.

16.2 EXCAVATION OF THE TUNNELS AND STRAIN DISTRIBUTIONS AROUND THEM

Excavation was simulated by reducing the airbag pressure systematically until collapse of the tunnels occurred. At each airbag pressure reduction, the surface settlements, the crown convergence and the incremental stress were measured. A photograph of the setup was also taken before the next decrease of the airbag pressure. The nodal displacements at the different airbag pressure levels were obtained by digitising the photographs using a Graphtec Digitizer KD 5000. The strain distribution around the tunnels can be calculated from measured displacements caused by the excavation (reducing air pressure) by considering the kinematic relationship between strains and displacements as shown in Equation (16.1).

$$\{\varepsilon\} = [B]\{u\} \tag{16.1}$$

where $\{\varepsilon\}$: strain around the tunnels; $\{u\}$: measured displacements at measuring points; $[B]$: matrix of the relationship between strain and displacement. The maximum shear strain, γ_{max}, is then derived by the following equation:

$$\gamma_{max} = |\varepsilon_1 - \varepsilon_3| \tag{16.2}$$

where ε_1 and ε_3 are the maximum and minimum principal strains, respectively.
One of the experimental results is shown in Figure 16.3.

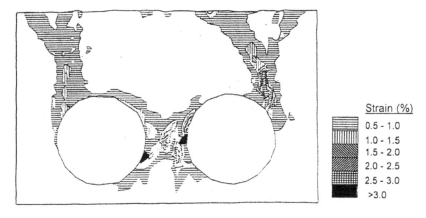

Strain (%)
0.5 - 1.0
1.0 - 1.5
1.5 - 2.0
2.0 - 2.5
2.5 - 3.0
>3.0

Figure 16.3 Maximum shear strain distribution around tunnels (experimental result).

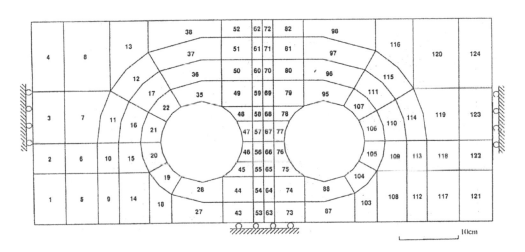

Figure 16.4 Mesh used for finite element analysis together with element numbers.

16.3   BACK ANALYSIS FOR SIMULATING THE MAXIMUM SHEAR STRAIN DISTRIBUTIONS

16.3.1   Optimisation of anisotropic parameter

In order to verify the applicability of the constitutive equation proposed in Section 14.4 for back analyses, measured displacements around the two parallel tunnels in the physical model tests were used as input data for back analyses. The constitutive equation shown in Equation (14.19) is used, in which anisotropic parameter, *m*, plays an important role. Since the back analysis procedure is performed with a finite element method, its element mesh, together with element numbers of the two-tunnel system, is shown in Figure 16.4.

Table 16.1 Mechanical parameters used in back analysis.

Unit weight	W	1.8 kNm^{-3}
Young's modulus	E	540.5 kPa
Poisson's ratio	v	0.33
Internal friction angle	ϕ	28.8°

Table 16.2 Back-analysed values of m before collapse of tunnels.

Left shear band			Centre			Right shear band		
Element	m values	*θ°	Element	m values	θ°	Element	m values	θ°
35	0.0007	8.66	59	0.10	9.25	79	0.05	9.10
36	0.0009	7.69	60	0.12	9.10	80	0.07	9.23
37	0.001	7.33	61	0.14	10.30	81	0.09	9.26
38	0.003	7.94	62	0.16	8.79	82	0.11	9.44
49	0.10	9.30	69	0.05	9.02	95	0.00009	9.61
50	0.12	9.32	70	0.07	9.06	96	0.0002	9.85
51	0.14	9.10	71	0.09	10.1	97	0.0004	9.40
52	0.16	9.95	72	0.11	9.98	98	0.0006	9.23

*Tabulated θ values are θ_l values

The objective of the back analysis is to determine the values of the anisotropic parameter that correctly simulate the deformational behaviour observed in the physical model tests. In order to achieve this, proper values of the anisotropic parameter can be determined to minimise the error function given in Equation (2.3). When the error function is minimised, the anisotropic parameter (m) values are optimised. In the back analyses, in addition to the m values, the mechanical parameters shown in Table 16.1 are used (Akayuli & Sakurai, 1998).

16.3.2   Minimisation of the error function

The minimisation of the error function is an iterative procedure leading to the optimisation of the m values at any particular airbag pressure. At each airbag pressure reduction, values of m and the material properties are input into the finite element code, and the incremental displacements are calculated. These incremental displacements, and the measured relative displacements obtained from the physical model test at the same airbag pressure, are input into the error function shown in Equation (2.3) and the value of δ is calculated.

Without changing the airbag pressure, a new set of m values is input into the finite element code, while keeping the other mechanical parameters constant. New incremental displacements are then calculated. The incremental displacements are again input into the error function and a new δ, say δ^j, is calculated. This calculation is repeated until the difference between δ^j and δ^{j-1} becomes sufficiently small to indicate that the error function has been minimised, resulting in the values of m being optimised for investigation of the stability of pillars in the room-and-pillar excavation method in underground mines.

In Table 16.2, the back-calculated values of m are shown. These values were then implemented in the $[D]$ matrix for use in the calculation of displacements, strains and stresses for each element.

16.4 RESULTS AND DISCUSSIONS

The maximum shear strain distribution obtained by the proposed back analysis method is shown in Figure 16.5. The results of the back analyses were compared with the experimental results shown in Figure 16.3, and it is obvious that they indicate good agreement with each other. The shear bands clearly extend from the shoulder of the tunnels to the surface of the model. Moreover, the deformation of the pillar zone is well simulated with that observed in the laboratory model test. Considering these results, it can be concluded that back analysis based on the proposed constitutive equation is highly applicable to practical problems.

For further reference, a conventional elasto-plastic analysis is carried out assuming perfectly elasto-plastic materials. In the analysis, the generalised Drucker-Prager yield criterion under an associated flow rule is assumed. The material properties used in the elasto-plastic analysis are given in Table 16.2; the cohesion of the materials was $c = 0.26\,\text{kPa}$. The maximum shear strain distribution obtained by the elasto-plastic analysis is given in Figure 16.6, which is entirely different from the result obtained by the physical model test shown in Figure 16.3. This difference may be due to the fact that the octahedral shear stress is taken into account in the Drucker-Prager yield criterion, as already described, with the result that no failure planes can be simulated.

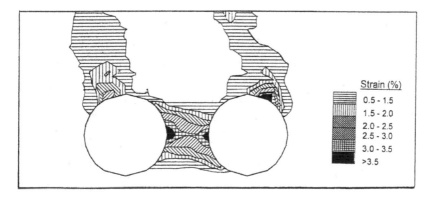

Figure 16.5 Maximum shear strain distribution around two parallel tunnels (back-calculated results).

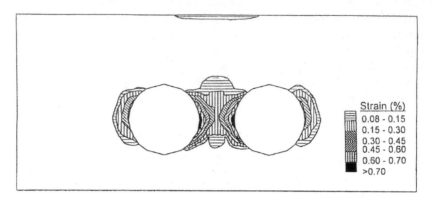

Figure 16.6 Maximum shear strain distribution around two tunnels (elasto-plastic analysis).

In conclusion, if we have sufficient numbers of measurement data, very precise strain distributions around structures can be determined. Once the strain distributions are back-calculated, the maximum shear strains are easily calculated, and comparing the back-calculated maximum shear strains with the critical shear strain of the materials involved, the stability of the structures can be easily assessed during construction.

Chapter 17

Assessing the stability of slopes

17.1 FACTOR OF SAFETY OF SLOPES

It should be noted that the mechanism of slope failure is usually assumed as sliding along a slip plane even though various shapes of slip planes exist, and the Factor of Safety (Fs) is usually defined as the ratio of resistance force (shear strength of geo-materials) to sliding force (mainly due to gravitational force) along a slip plane, as shown in Equation (17.1). A slope is designed so as to make the factor of safety always greater than an allowable value; usually, the allowable value is greater than 1.2 in slope engineering practice.

$$Fs = \frac{\sum R_i}{\sum T_i} \geq 1.2 \tag{17.1}$$

where $\sum R_i$: resistance force (shear strength of geomaterials); $\sum T_i$: sliding force along a sliding plane; $R_i = f(c, \varphi \dots)$.

In the design of slopes, the strength parameters of the geomaterials, such as cohesion and internal friction angle, must first be evaluated. However, it is not an easy task to evaluate these strength parameters, particularly for highly jointed rock masses, because of their complex geological and geomechanical characteristics. Consequently, the accuracy and reliability of the strength parameters becomes questionable. To overcome this difficulty, field measurements are carried out during the excavation of slopes in difficult geological and geomechanical ground. However, the question may arise of how to determine the strength parameters from the field measurement results. Answering the question, Sakurai (1993) proposed a back analysis procedure to determine the strength parameters of geomaterials from measured displacements.

Among the various kinds of field measurements, displacement measurements are most commonly performed because of their reliability and ease of handling. However, displacements are not generally taken into account in the design of slopes, with the result that there is no way to evaluate the measured displacements. Furthermore, practising engineers are aware of the displacements of slopes during construction sometimes being large, so that additional support measures are needed. However, there is no standard criterion as to when such additional support measures must be installed. As a result, the decision about when they are installed to stabilise a slope is made on the judgement of the individual engineer.

17.2   PARADOX IN THE DESIGN AND MONITORING OF SLOPES

The design of slopes is usually performed on the basis of the factor of safety, which is determined by comparing the sliding force to the resistance force acting along a slip plane, as shown in Equation (17.1). In this design approach, displacements are not taken into account. In monitoring, however, displacements are usually measured to assess the stability of a slope. Therefore, it is questionable as to how the validity of the original design of a slope can be confirmed from measured displacements. Nevertheless, in the monitoring of slope stability, the magnitude of both measured displacements and their rate of change are often used as criteria for assessing slope stability, even though they are only empirical criteria based on an engineer's experiences. Thus, in the design of slopes the factor of safety is used, which is defined in terms of stress, while in monitoring slopes, measured displacements together with the changing rate of displacements are used, even though there is no linkage between the two. This is a paradox of the design and monitoring of slopes in engineering practice.

In order to overcome this paradox, Sakurai and Nakayama (1999) proposed a back analysis procedure to determine the strength parameters of the geomaterials, such as cohesion and the internal friction angle, from measured displacements, with the result that the factor of safety can be calculated and compared with the one used at the design stage to assess the adequacy of the original design (see Section 21.2). This back analysis can create a bridge between the design and monitoring of slopes; in other words, the measured displacements can be closely linked to the slope design, even though the displacements were not considered at the design stage. It should be noted, therefore, that the validity of the original design of a slope can be confirmed by using strength parameters back-calculated from displacements measured during its construction.

In order to link the design of slopes to the monitoring of them, two approaches for assessing the stability of slopes are available in terms of the factor of safety. The first approach is a stress-based approach, in which the maximum shear stress back-calculated from measured displacements is compared with the critical shear stress of the geomaterials. By contrast, the second approach is a strain-based approach, in which the maximum shear strains occurring in the slopes can be determined from measured displacements with the use of the kinematic relationship between displacements and strains, as described in Section 7.3, with the result that the stability of a slope can be assessed by comparing the critical shear strain of its geomaterials to the maximum shear strains occurring in the slope. In this second approach, the number of measurement data is usually not large enough to determine the maximum shear strain distributions directly from measured displacements by using the kinematic relationship. If this is the case, a back analysis procedure can be used to determine the maximum shear strains from a limited number of measured displacements.

17.3   DIFFERENCE BETWEEN THE FACTOR OF SAFETY OF TUNNELS AND SLOPES

17.3.1   Tunnels

As already described in Section 4.3, the stability of tunnels cannot be assessed by the factor of safety defined in terms of stress. This is due to the fact that tunnels are structures which only allow the geomaterials (surrounding rocks) to be used in a plastic state. The results is that in the plastic state the factor of safety defined in terms of stress always becomes $Fs = 1.0$, even though the structures are still stable. To solve this problem, the strain-based approach may be promising for assessing the stability of tunnels, where the factor of safety is defined as the ratio of the critical shear strain of the rock mass to the maximum shear strain occurring in the rock mass, as shown in Equation (17.2).

$$Fs = \frac{\gamma_0}{\gamma_{max}} \tag{17.2}$$

where γ_0: critical shear strain of geomaterials; γ_{max}: maximum shear strain occurring in the rock mass.

It should be noted that the factor of safety defined in Equation (17.2) is not for assessing the overall stability of tunnels, but for assessing the stability of a small zone around the tunnels. Therefore, even if the factor of safety falls below unity ($Fs < 1.0$), overall failure is not necessarily occurring. In addition, the critical shear strain of geomaterials varies widely, as shown in Figure 7.2, meaning that it may be possible for the factor of safety defined by Equation (17.2) to be less than one ($Fs < 1.0$) even though the tunnels are still stable. In order to overcome these problems, hazard warning levels are proposed according to the relationship between the maximum shear strain and shear modulus of geomaterials around the tunnels, as shown in Figure 7.5, which is extended from the hazard warning levels expressed as the relationship between the uniaxial compressive strain and uniaxial compressive strength of geomaterials, shown in Figure 5.8 (Sakurai, 1997a).

17.3.2 Slopes

Regarding the factor of safety of slopes, the stress-based approach given in Equation (17.1) is popularly used for assessing the stability of slopes in terms of sliding. This is simply because it is easily understood by geotechnical engineers that if the sliding force (calculated by the gravitational force) along a slip plane becomes the same as the resistance force (evaluated by the shear strength of the geomaterials), the factor of safety becomes equal to one ($Fs = 1.0$), as seen in Equation (17.1), with the result that failure may occur. It is noted that sliding is the major failure mechanism of slopes consisting of soils, while in rock slopes many different types of failure exist depending on complex jointed rock systems. Goodman & Kieffer (2000) classified rock slope hazards in two groups, i.e. rock slumping and toppling, and summarized the failure of rock slopes according to various types of failure modes.

It should be noted that for assessing the stability of the toppling of slopes, it is questionable whether we can apply Equation (17.1) because no slip plane exists. Thus, it might seem possible to use Equation (17.2) (strain-based approach) for assessing the stability of toppling instead. However, as already described, the factor of safety defined in Equation (17.2) cannot be used to evaluate the overall stability of slopes. In addition, the factor of safety determined by the strain approach is not popularly used in slope engineering practice. Given these circumstances, the stress-based approach may be preferable even for toppling of slopes, in such a way that the strength parameters of the geomaterials can be back-calculated from the measured displacements, with the result that the factor of safety of slopes can be calculated using Equation (17.1), which can assess the overall stability of slopes (Sakurai et al., 2009).

17.4 FACTOR OF SAFETY FOR TOPPLING OF SLOPES

In jointed rock masses, toppling failure often occurs, so that the stability of toppling failure must be properly assessed. It is obvious that the failure mechanism of toppling is not sliding, but may be the collapse of jointed rock masses associated with rotation of block movements. Therefore, the conventional (stress-based) approach cannot be applied for determining the factor of safety, because no particular slip plane exists. Toppling is a specific characteristic of highly jointed rock masses, and it has never appeared in soils. An illustration of toppling is provided in Figure 17.1 (Goodman & Bray, 1976).

In order to assess the toppling failure of jointed rock masses, the discrete (distinct) element method (DEM) (Cundall, 1977) may be a potential tool, provided that we know all the joint systems. However, it is almost impossible to explore all of the joint systems of rock masses. As an alternative way, a probabilistic approach may be used for identifying the joint systems from the field observation data. The probabilistic approach seems to be promising, but it is hardly applicable to engineering practice, because not many data are usually available. Considering the shortcomings

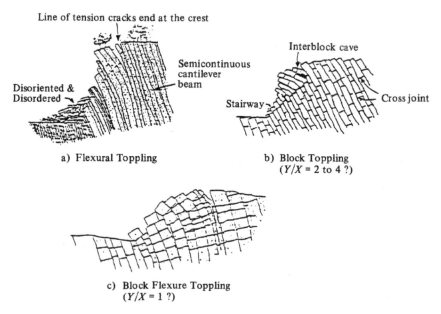

Figure 17.1 Illustration of toppling behaviour of jointed rocks (Goodman & Bray, 1976, with some additions).

of the probabilistic approach, a deterministic approach may be preferable, particularly for practising engineers. Moreover, the stress-based approach determining the factor of safety by Equation (17.1) may be advantageous, because it can assess the overall stability for toppling failures. The detail of the stress-based approach is described in Section 21.2.

Chapter 18

Back analysis of slopes based on the anisotropic parameter

18.1 MECHANICAL MODEL OF ROCK MASSES

Rock masses are generally classified into three groups as shown in Figure 18.1. Class A is a continuous type, class B is a discontinuous type, and class C is a highly fractured rock type. The overall behaviour of rock mass of class C may be most likely to be a continuous material, so that it is called the pseudo-continuous type.

The class A type is not only representative of soils, but also of weak rocks where the matrix of the materials is so weak that the effect of joints does not influence much of the overall behaviour of the rocks. Thus, the mechanical behaviour of slopes of the class A type can be analysed by using a continuum mechanics approach, that is, a soil mechanics approach. For the class B type, a discontinuous model such as those proposed by Cundall (1971) and Kawai (1980) can be used. For this class of rocks the finite element method with joint elements can also be used (Goodman et al., 1968).

On the other hand, for the class C type of rocks, a discontinuous model can be used, similar to that of class B. In engineering practice, however, it is almost impossible to explore all of the joint systems together with their mechanical characteristics. Moreover, it seems that in an overall view of this type of rock it behaves like a continuous material, with the result that a continuous mechanics model may be adopted for analysing the stability of highly jointed rock masses. In addition, in engineering practice a continuum mechanics approach must be superior to the discrete modelling method, because numerical analysis must be simple enough to be able to feedback the back analysis results to engineering practice immediately after gathering the data.

Nevertheless, for the class B type special consideration must be paid to taking into account the effects of discontinuities on the mechanical behaviour of rock masses. In this book, only the continuous and pseudo-continuous types of rocks are described.

For the continuous and pseudo-continuous types of rock, the deformational behaviours of slopes are classified into three different modes, that is, elastic, sliding and toppling, as shown in Figure 18.2. Toppling is the specific characteristic of the pseudo-continuous type of rocks. This type of mode has never appeared in soils; it occurs only in highly jointed rock masses.

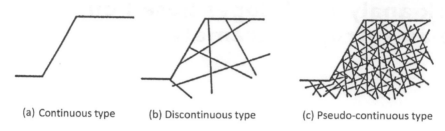

(a) Continuous type   (b) Discontinuous type   (c) Pseudo-continuous type

Figure 18.1 Classification of jointed rocks.

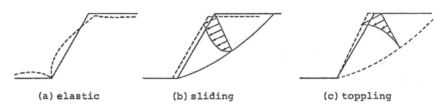

(a) elastic   (b) sliding   (c) toppling

Figure 18.2 Deformational modes of slopes (continuous and pseudo-continuous types of rock).

18.2   LABORATORY EXPERIMENTS FOR TOPPLING

In the forward analyses of slopes at the design stage, we must first assume a mechanical model for deformational modes, choosing one from among the three modes, such as elastic, sliding or toppling, by considering the results of geological and geomechanical field explorations. Among these three deformational modes, toppling is the most difficult to mechanically model. In order to investigate the toppling mechanism of slopes, physical model tests were performed in the laboratory, in which two-dimensional block models, consisting of a large number of aluminium bars of 1 cm × 1 cm in cross section and 5 cm in length were piled up in different joint inclinations to simulate jointed rock masses (see Figures 18.3 and 18.4) (Deeswasmongkol & Sakurai, 1985, 1986).

The platform of the model is movable in order to pile the aluminium bars at various joint angles. A sketch of the physical model is shown in Figure 18.3. Before starting the experiments, the surface of the model is horizontal, and the excavation is performed by hand, layer by layer.

From careful investigations of the results of the physical model tests, it has been found that the shear deformation of the material above the base line takes place easily, and the largest shear deformation occurs almost parallel to the direction of the cross joints, as shown in Figure 18.4. The base line is defined as a line under which no displacement occurs at all, and it may be slightly steeper than the angle of the cross joints.

The displacements are measured during the removal of some of the blocks, step by step, to simulate the cutting of slopes. The measured displacement vectors are shown in Figure 18.5. It is obvious from the figure that a toppling deformation occurs due to cutting of the slope.

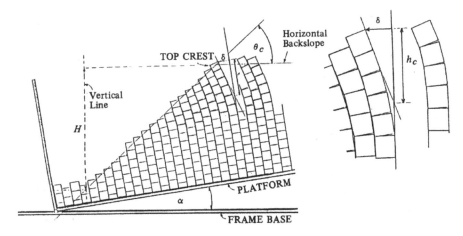

Figure 18.3 Sketch of the block model used to investigate toppling.

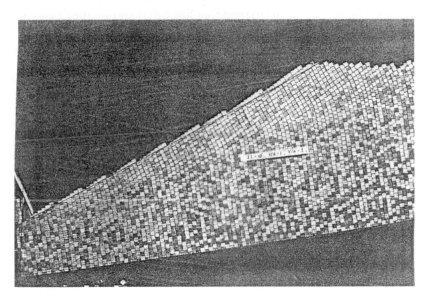

Figure 18.4 Physical model consisting of a large number of aluminium blocks to simulate cut slope (toppling behaviour) (Deeswasmongkol & Sakurai, 1985, 1986; Sakurai et al., 1986).

18.3 NUMERICAL ANALYSIS OF TOPPLING BEHAVIOURS

18.3.1 Introduction

In numerical simulations it may be possible to analyse toppling behaviours by using the discrete element method (DEM). As already described, however, the DEM is only applicable when all of the joint systems of rock masses are known. It is obvious that it is impossible in engineering practice to identify all of the joint systems at the design stage.

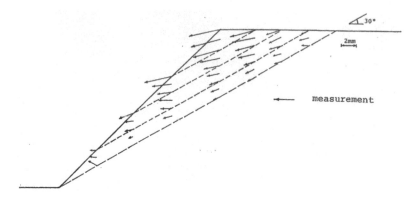

Figure 18.5 Displacement distribution of slope due to excavation (results of laboratory experiment).

In addition, as far as the design approach to toppling behaviours is concerned, no text-book is available that describes how to treat toppling behaviours in the design of slopes.

18.3.2   Constitutive equation

In a back analysis, toppling behaviours can be treated numerically in the same manner as the other two types of behaviours, provided that the anisotropic parameter is used.

In order to simulate the results of the physical model tests on the deformational behaviour of jointed materials, the anisotropic parameter is used (Sakurai et al., 1986). Even though the constitutive equation used for the simulation of model test results has already been described in Section 14.4, it is presented again here.

Let the local coordinate system $x' - y'$ be taken as shown in Figure 18.6 where the x'-axis is parallel to the direction of the cross joints. Then the constitutive equation considering the anisotropic parameter can be described as follows (Sakurai, 1987):

$$\left\{ \begin{array}{c} \sigma_{x'} \\ \sigma_{y'} \\ \tau_{x'y'} \end{array} \right\} = [D'] \left\{ \begin{array}{c} \varepsilon_{x'} \\ \varepsilon_{y'} \\ \gamma_{x'y'} \end{array} \right\} \tag{18.1}$$

where

$$[D'] = \frac{E}{1 - \nu - 2\nu^2} \begin{bmatrix} 1 - \nu & \nu & 0 \\ \nu & 1 - \nu & 0 \\ 0 & 0 & m(1 - \nu - 2\nu^2) \end{bmatrix}$$

where E: Young's modulus
   ν: Poisson's ratio
   m: anisotropic parameter.

Equation (18.1) is valid in the local coordinate system. The x'-axis is taken to be parallel to a potential slip plane where the stress state satisfies the Mohr–Coulomb failure criterion.

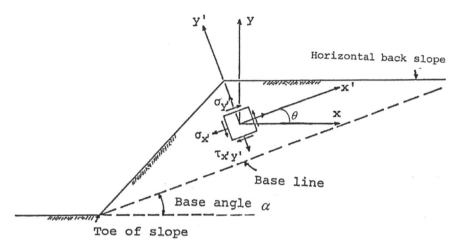

Figure 18.6 Local and global coordinate systems.

When the constitutive equation for the local coordinate system is known, it is easily extended to the x–y global coordinate system, as shown in Equations (18.2) to (18.3).

$$
\begin{Bmatrix} \sigma_x \\ \sigma_y \\ \tau_{xy} \end{Bmatrix} = [D] \begin{Bmatrix} \varepsilon_x \\ \varepsilon_y \\ \gamma_{xy} \end{Bmatrix}
\tag{18.2}
$$

where

$$
[D] = [T][D'][T]^T
\tag{18.3}
$$

where $[T]$ is a transformation matrix given as a function of the angle θ between the x- and x'-axes, as shown in Figure 18.6.

$$
[T] = \begin{bmatrix} \cos^2 \theta & \sin^2 \theta & -2 \sin \theta \cos \theta \\ \sin^2 \theta & \cos^2 \theta & 2 \sin \theta \cos \theta \\ \sin \theta \cos \theta & -\sin \theta \cos \theta & \cos^2 \theta - \sin^2 \theta \end{bmatrix}
\tag{18.4}
$$

18.3.3 Mechanical model of slopes

In the analysis of cut slope problems, the material above the base line is divided into N layered elements of zones parallel to the base line, as shown in Figure 18.7. The model may be extended to more general slope problems with a curved failure plane using curved layers. It is assumed that each layer has a different value of m, but the same value of α, E and v. It is also assumed that the material below the base line behaves as a homogeneous, isotropic elastic material, so that only two material constants (E_0, v_0) exist.

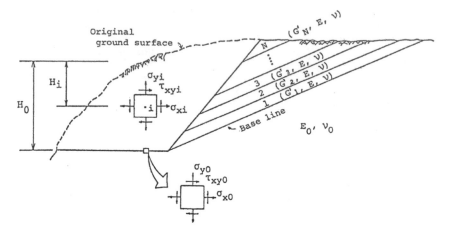

Figure 18.7 Layered elements of zones parallel to the base line.

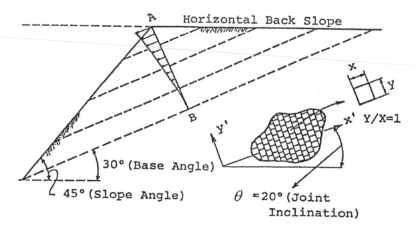

Figure 18.8 Mechanical model and measured displacements.

18.3.4 Applicability of the back analysis method to toppling behaviours

In order to verify the applicability of the back analysis method to the toppling behaviour of slopes, the method is applied to the analysis of the displacements measured in the physical model study described in Section 18.2. The medium above the base line is divided into four layers, as shown in Figure 18.8, and the anisotropic parameters, m_1, m_2, m_3 and m_4, together with the normalised initial stresses σ_{x0}/E, σ_{y0}/E and τ_{xyo}/E, can be back-calculated. The base angle is assumed to be $\alpha = 30°$ on the basis of the results of the physical model study.

The displacements along line A–B of the model, shown in Figure 18.8, are taken to be the measured displacements which are used as input data for the back analysis.

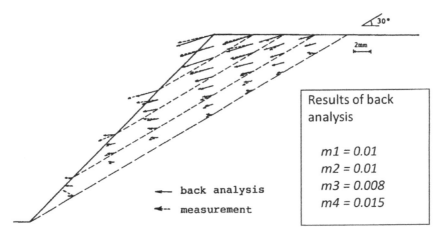

Figure 18.9 Comparison between measured and back-calculated displacements using anisotropic parameters.

The measured displacements shown in Figure 18.5 are used as the input data for the back analysis. As a result, the anisotropic parameters, together with normalised initial stresses, are back-calculated so as to achieve a good agreement between the measured and the back-calculated displacements. In this calculation, the error function, given in Equation (2.3), should be minimised.

The results of the back analyses for the anisotropic parameters and normalised initial stresses are given as follows:

$m_1 = 0.01$ $\sigma_{x0}/E_R = -2.15 \times 10^{-3}$
$m_2 = 0.01$ $\sigma_{y0}/E_R = -0.60 \times 10^{-3}$
$m_3 = 0.008$ $\tau_{xy0}/E_R = 0.0$
$m_4 = 0.015$ (tensile stress is positive)
Angle of layer $= 20°$

Once the anisotropic parameters and the normalised initial stresses are back-calculated, the displacement distribution can be determined by forward analysis, using the back-calculated anisotropic parameters and the normalised initial stresses as input data. The results of the displacement vectors are shown in Figure 18.9, where the measured displacements are also shown. It can be seen that both the back-calculated and measured displacements coincide well with each other.

It is of interest to know that the toppling deformation of such a discontinuous material consisting of a large number of aluminium bars can be successfully back-calculated from measured displacements by using the constitutive equation, Equation (14.19), which is formulated on the basis of a continuum mechanics approach using the anisotropic parameters. Moreover, it is surprising to find that in this model study just four values of anisotropic parameter can work well to simulate a toppling deformation of the jointed material. In addition, it should be emphasised that if all four numbers

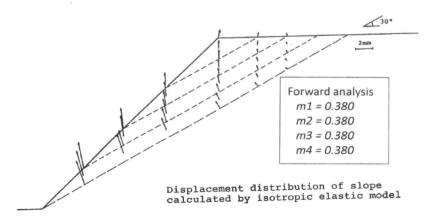

Forward analysis
m1 = 0.380
m2 = 0.380
m3 = 0.380
m4 = 0.380

Displacement distribution of slope
calculated by isotropic elastic model

Figure 18.10 Displacement distribution of slope calculated by isotropic linear elastic model.

are assumed to be the same as $m = 0.380$ (Poisson's ratio: $v = 0.3$), then the overall displacement distribution becomes that of an isotropic linear elastic material, as shown in Figure 18.10. It is obvious from Figure 18.10 that all the displacement vectors are oriented upward due to the excavation, that is, rebound occurs. There exists a large difference between the measured displacements (Figure 18.5) and those for isotropic linear elastic materials (Figure 18.10).

It should be noted that if the four values of anisotropic parameter differ from each other, not only toppling but also sliding behaviour can be simulated, as shown in Section 18.4.1.

18.4   APPLICABILITY OF THE ANISOTROPIC PARAMETER TO SIMULATION OF VARIOUS DEFORMATIONAL MODES

18.4.1   Three different deformational modes

It can be demonstrated that three different types of deformational mode of slopes, such as elasticity, sliding and toppling, are simulated by a forward analysis with the use of the identical constitutive equation, that is, Equation (14.19), by changing only the input data of the anisotropic parameters under a given Young's modulus (Poisson's ratio: $v = 0.3$) and the initial stresses. The finite element method is used for the simulation, where the medium above the base line is divided into four layers in the same fashion as for the toppling shown in Figure 18.8. In each layer, the anisotropic parameter, m_1, m_2, m_3 and m_4, respectively, varies, and the base line angle is assumed to be $\alpha = 30°$, as shown in Figure 18.11. All the input data for the forward analysis are shown in Table 18.1.

The results of the simulation are given in Figure 18.12, which shows the displacement distributions, determined by using four different values for the anisotropic parameters in the layers (Sakurai, 1990).

It is obvious from Figure 18.12 that the anisotropic parameters are very applicable to simulation of the three different deformational modes. This implies that it may

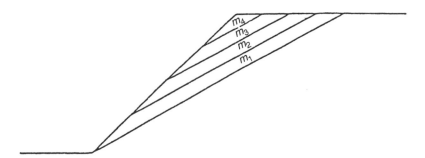

Figure 18.11 Four layers above the base line.

Table 18.1 Input data used in the numerical simulations of slopes.

	Elastic	Sliding	Toppling
E (MPa)	500.0	100.0	200.0
v	0.3	0.3	0.3
m_1	0.385	0.005	0.020
m_2	0.385	0.385	0.010
m_3	0.385	0.385	0.008
m_4	0.385	0.385	0.005
σ_{x0} (MPa)	−0.16	−1.08	−1.08
σ_{y0} (MPa)	−0.38	−0.30	−0.30
τ_{xy0} (MPa)	0.11	0.0	0.0

be possible to back-calculate the deformational mode from measured displacements without considering any mechanical model, and only through changing the values of the anisotropic parameters. This is a great advantage for the back analysis, inasmuch that the mechanical model is not necessarily assumed, but the deformational modes can be identified by back analyses of the measured displacements. Furthermore, the back analysis for determining the anisotropic parameters may be achievable from the surface displacements alone. This is demonstrated by a numerical example provided in Section 18.4.2.

18.4.2 Monitoring slope stabilities by displacements measured on the ground surface

18.4.2.1 Introduction

In the monitoring of slopes, conventional measurement systems, such as extensometers, borehole inclinometers, total stations and laser displacement meters, are barely applicable, because they are limited to measuring ground movements only in a small area. In the case of measuring displacements of the ground over a large extent, GPS displacement measurements provide an effective tool. However, GPS can only measure the surface displacements of the ground, and thus we need some technique to

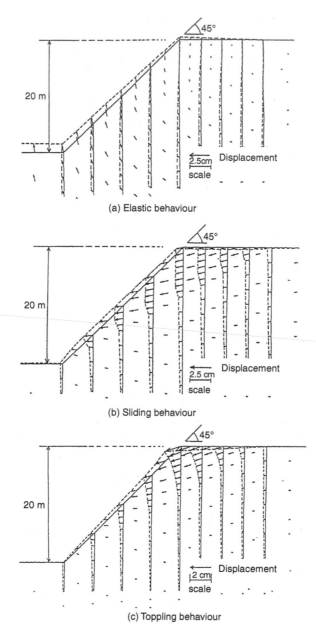

Figure 18.12 Three different deformational modes (numerical simulations) (Sakurai, 1990).

evaluate the deformational behaviour inside the ground from these measured surface displacements.

Displacements of slopes occur due to either excavations of the ground or a decrease in the shear strength of geomaterials. The latter mechanism equates to landslides, which may be caused by a decrease in shear strength; in other words, the anisotropic parameter, m, decreases in a certain zone inside the ground. In landslides, various types of deformational mechanism may exist, such as sliding with a straight slip plane, sliding with a curved slip plane, surface sliding, and toppling, and, furthermore, these may either be at shallow depth or at great depth. In considering these deformational mechanisms of landslides, the relationship between the surface displacements and the deformational behaviours inside the ground must be investigated. For this purpose, simple numerical simulations are carried out using the anisotropic parameters.

18.4.2.2 Numerical simulations on deformational modes of slopes

Numerical simulations (forward analyses) are carried out to demonstrate the deformational modes of landslides that are associated with different values of the anisotropic parameters. Because the anisotropic parameters are a key aspect of the simulations, the constitutive equation of geomaterials given in Equation (14.19) is adopted. In the numerical simulations, it is assumed that the deformation of slopes is caused by a reduction in the anisotropic parameter, where no excavation activities are operating (Sakurai & Hamada, 1996). The detail of the mathematical formulation of the finite element analysis for landslides is described in Section 20.2.

In the numerical simulations, the finite element method is used, and the anisotropic parameters in each subdivided region are assumed for the input data, so as to simulate each deformational mode, such as (a) sliding (straight sliding plane), (b) sliding (curved sliding plane), (c) surface sliding and (d) toppling, by a trial and error method. In the simulations, Young's modulus, Poisson's ratio ($v = 0.3$) and the base angle α are assumed. All the input data used in the simulations (forward analyses) for each deformational mode, together with the finite element mesh, are shown in Figure 18.13.

The simulation results are also shown in Figure 18.13, where displacement vectors on the ground surface are indicated for each deformational mode, together with the anisotropic parameters used as input data in each subdivided region. These results show that the displacement distributions on the ground surface are closely related to the location of damaged zones where the anisotropic parameters decrease. The damaged zone is indicated by broken lines for the cases (a), (b) and (c), while for the case (d), all the four zones above the base line are damaged zones. It can be seen from these figures that there is a possibility of developing some technique for identifying the type of deformational mode from the displacements measured on the ground surface. Thus, it may be possible to predict the location and size of damaged zones from the surface displacements measured by GPS, provided that a sufficient amount of GPS data is available. In addition, if some displacements inside the ground are measured by borehole inclinometers, of course, the accuracy of back-calculation for the location as well as the size of damaged zones must be increased.

18.5 FACTOR OF SAFETY BACK-CALCULATED FROM MEASURED DISPLACEMENTS

The factor of safety can be determined in two different ways: one is a strain-based approach and the other is stress-based. In the strain-based approach, the factor of safety can be defined as the ratio of the critical shear strain of geomaterials to the

(a) Sliding (straight sliding plane)

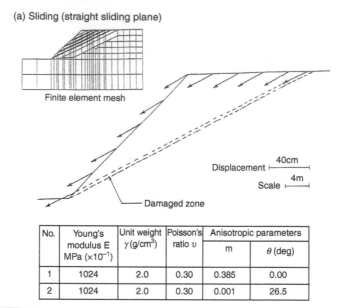

Finite element mesh

Displacement ⊢————⊣ 40cm

Scale ⊢——⊣ 4m

Damaged zone

No.	Young's modulus E MPa (×10⁻¹)	Unit weight γ (g/cm³)	Poisson's ratio υ	Anisotropic parameters	
				m	θ (deg)
1	1024	2.0	0.30	0.385	0.00
2	1024	2.0	0.30	0.001	26.5

(b) Sliding (curved sliding plane)

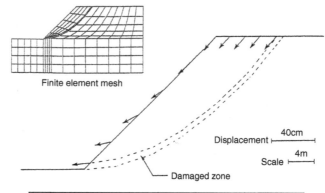

Finite element mesh

Displacement ⊢————⊣ 40cm

Scale ⊢——⊣ 4m

Damaged zone

No.	Young's modulus E MPa (×10⁻¹)	Unit weight γ (g/cm³)	Poisson's ratio υ	Anisotropic parameters	
				m	θ (deg)
1	1024	2.0	0.30	0.385	0.00
2	1024	2.0	0.30	0.001	6.39
3	1024	2.0	0.30	0.001	18.37
4	1024	2.0	0.30	0.001	29.07
5	1024	2.0	0.30	0.001	37.88
6	1024	2.0	0.30	0.001	45.00
7	1024	2.0	0.30	0.001	50.71

Figure 18.13 Displacement vectors on the ground surface due to different mechanisms of deformation (Sakurai & Hamada, 1996).

(c) Surface sliding

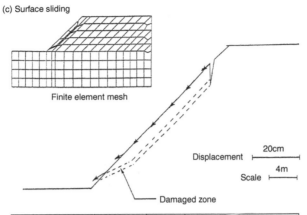

Finite element mesh

Displacement ⊢————⊣ 20cm

Scale ⊢——⊣ 4m

Damaged zone

No.	Young's modulus E MPa ($\times 10^{-1}$)	Unit weight γ (g/cm^3)	Poisson's ratio υ	Anisotropic parameters	
				m	θ (deg)
1	1024	2.0	0.30	0.385	0.00
2	1024	2.0	0.30	0.001	24.00
3	1024	2.0	0.30	0.001	45.00
4	1024	2.0	0.30	0.001	45.00
5	1024	2.0	0.30	0.001	45.00

(d) Toppling

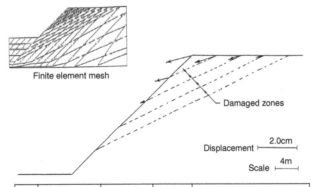

Finite element mesh

Damaged zones

Displacement ⊢————⊣ 2.0cm

Scale ⊢——⊣ 4m

No.	Young's modulus E MPa ($\times 10^{-1}$)	Unit weight γ (g/cm^3)	Poisson's ratio υ	Anisotropic parameters	
				m	θ (deg)
1	1024	2.0	0.30	0.385	0.00
2	1024	2.0	0.30	0.005	15.00
3	1024	2.0	0.30	0.01	15.00
4	1024	2.0	0.30	0.05	15.00
5	1024	2.0	0.30	0.1	15.00

Figure 18.13 (Continued).

maximum shear strain occurring in the ground of the slope, as shown in Equation (17.2). As already described in Section 17.3, however, the strain-based approach has the disadvantage that it cannot evaluate the overall stability of a slope, because the factor of safety determined by the strain-based approach is for assessing a local failure at the point where the damaged zone appears in the ground of the slope.

Given this drawback of the strain-based approach, the stress-based approach defined by Equation (17.1) would seem preferable, and it is popularly adopted in slope engineering practice, because it is possible to evaluate the factor of safety for the overall stability of a slope. In the stress-based approach, both the resistance force and the sliding force acting on a sliding block of rock masses above a slip plane are considered in evaluating the stability of the slope, in such a way that if the sliding force is greater than the resistance force, the factor of safety becomes less than or equal to one ($FS \leq 1.0$), resulting in slope failure. It must be clearly understood that the sliding force is mainly caused by gravitational force (external force), which is proportional to the volume of a sliding block, while in the strain-based approach, the maximum shear strains are considered but they are not directly related to gravitational force. This means that it is difficult to evaluate the effect of gravitational force in terms of strains.

On the other hand, the resistance force for the sliding blocks is evaluated by the shear strength of geomaterials in the damaged zone. However, it is not easy to determine the shear strength of geomaterials from measured displacements. To overcome this difficulty, Sakurai and Nakayama (1999) proposed a back analysis procedure to back-calculate the shear strength from measured displacements. Once the strength parameters are back-calculated, the factor of safety of slopes in terms of overall stability is calculated by the conventional limit equilibrium method, such as the Fellenius method, shown in Equation (21.5) (Sakurai et al., 2009). In the proposed back analysis procedure, the critical shear strain of geomaterials is a key to link the measured displacements to the shear strength of geomaterials.

The proposed back analysis approach has the great advantage that it is not necessary to assume any mechanical model for expressing the type of deformational mode, such as elasticity, sliding and toppling. This is simply due to the fact that the anisotropic parameter is introduced in the constitutive equation shown in Equation (14.19). Anisotropic parameters can work well to identify the damaged zones appearing in the ground of slopes. The damaged zone is defined as the zone where the anisotropic parameter decreases (see Section 18.4.2).

It should be noted that in toppling of jointed rock masses, it is obvious that there is no sliding plane, but there may exist a potential failure plane in the damaged zones, along which a toppling failure can occur. The potential failure plane must correspond to the base line defined in the physical model tests, described in Section 18.2.

It is of interest that even in the toppling of jointed rock masses it is possible to back-calculate the shear strength of geomaterials from measured displacements in the same fashion as the other deformational behaviours, such as elasticity and sliding, simply by changing the value of the anisotropic parameters. Once the anisotropic parameters have been back-calculated from the measured displacements, the shear strength of geomaterials can be determined, resulting in the factor of safety being calculated, as seen in Section 21.2.

Chapter 19

Back analysis method for predicting a sliding plane

19.1 INTRODUCTION

FEMs are commonly used in back analyses for interpreting measurement data. For finite element back analyses, the modelling of finite element meshes is a crucial problem because the accuracy and reliability of finite element analyses depend on what finite element meshes are used. In forward analyses, it must be difficult to create finite element meshes because there is no information about the deformational modes prior to construction, whereas in back analyses, some field measurement results are available so there is some information on the modelling of finite element meshes. In this context, the numerical simulation results shown in Section 18.4.2 provide very useful information, and reveal that it may be possible to identify deformational modes, such as elastic, sliding and toppling, by back analyses of displacements measured on the ground surface. If the deformational modes are evaluated from the measured displacements, the modelling of finite element meshes may be easily performed.

Considering simulation results, the baseline is important for creating the finite element meshes. In the case of sliding (curved slip plane) a curved baseline is required, but in other cases including toppling the baseline is a straight line, as seen in Figure 18.12, so that it may be easy to create the finite element meshes.

In the case of sliding, in particular, with a curved slip plane, it is important to evaluate the curved slip plane by back analysis of the surface displacements. Therefore, prior to back analysis, a sliding plane should be back-calculated from the displacements measured on the ground surface. For this purpose, a simple method for identifying a potential sliding plane from the measured surface displacements is proposed (Sakurai & Hamada, 1996). Once the potential sliding plane is determined by back analysis, finite element meshes can be easily created considering the back-calculated potential sliding plane. The proposed method is described in Section 19.2.

19.2 PROCEDURE OF THE METHOD

A two-dimensional case is illustrated here, although the method may be extended to a three-dimensional case. Now, let us consider the displacement vectors measured along the slope surface, as shown in Figure 19.1. It is assumed that the edge of the sliding plane – either point A or point D – is detected by field observation of the unusual deformational behaviours, such as cracks and swelling, appearing on the slope surface. If point A is detected as the edge of the sliding plane, a potential sliding

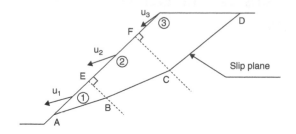

Figure 19.1 Schematic diagram for the proposed method.

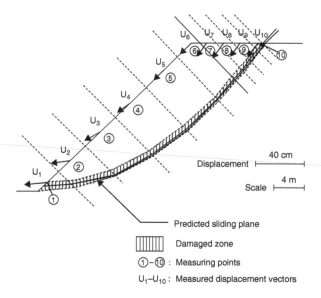

Figure 19.2 Application of the proposed method for predicting a sliding plane from the slope surface displacement vectors (Sakurai & Hamada, 1996).

plane starts from point A, parallel to displacement vector U_1 until hitting point B on a straight line perpendicular to the slope surface at point E located at the centre of the two measuring points, ① and ②. From point B, the sliding plane stretches parallel to displacement vector U_2 until hitting point C. After that, we repeat the same procedure until reaching the last point D. If some unusual deformational behaviours occur at point D, the predicted sliding plane is correct. If not, some compensation is needed in terms of reconsidering the location of points A and D.

19.3   ACCURACY OF THE METHOD

The accuracy of the proposed method for predicting a sliding plane is demonstrated by showing a numerical simulation using FEM. In this simulation, it is assumed that strain localisation has taken place along the sliding plane. As an example problem, the strain localisation zone (a shaded zone in the figure) is assumed to be a potential curved sliding plane, as shown in Figure 19.2. In other words, the shear rigidity in the shaded zone (damaged zone) decreases with an increase in shear strain, resulting in displacements on the slope surface being calculated by FEM. The calculated surface displacements are considered virtual 'measured displacements' that are used as input data for the back analyses proposed in Section 19.2. It should be noted that in this simulation the relationship between the anisotropic parameter and the maximum shear strain given in Equation (14.13) must be always satisfied during the iteration computation process.

Let us now suppose that these virtual measured displacements on the slope surface are used as input data such as $U_1–U_{10}$. A potential sliding plane can then be predicted by the proposed method. The results are also shown in Figure 19.2. It is obvious that the predicted sliding plane falls exactly within the shaded zone (damaged zone). This means that the proposed method can be applied to predicting a sliding plane from the surface measured displacements with high accuracy.

Chapter 20

Back analysis of landslides

20.1 INTRODUCTION

There are two causes of displacements in slopes. One is the change of stress distributions in the ground due to excavations and/or landfill. The other is the reduction of strength of geomaterials as time proceeds.

In finite element analyses for the excavation problem of underground openings, the genuine rock pressure (the initial tress) can be taken into account by applying equivalent nodal forces $\{P_0\}$ acting on the excavation surface, which corresponds to the initial state of stress $\{\sigma_0\}$ in the ground. This force is determined by

$$\{P_0\} = \int_V [B]^T \{\sigma_0\} dv + \int_V [N]^T \{p\} dv \tag{20.1}$$

where $\{p\}$ is the vector of the body force components due to gravity, $[N]$ and $[B]$ are the matrices of the element shape functions and their derivatives, respectively, and v is excavation volume.

In the case of the reduction of the strength of geomaterials, the anisotropic parameter m decreases, although there are no excavation activities. This type of deformation is caused by landslides, where the anisotropic parameter m decreases with increase in displacements caused by weathering and/or creep of geomaterials. However, it should be noted that no matter what the cause might be, the increase in shear strain causes the reduction of anisotropic parameter m (see Figure 20.1) (Sakurai, 1997b).

20.2 FINITE ELEMENT FORMULATION

The mathematical formulation based on FEM is described here. When the anisotropic parameter m decreases with a certain increment, say Δm (see Figure 20.1), the displacement increment $\Delta \delta$ is then related to Δm in the following equation:

$$[K]\{\Delta \delta\} = \{\Delta P\} \tag{20.2}$$

where $\{\Delta P\}$ is the external force acting at each nodal point, which is equivalent to the reduction of m values, and it can be represented by the following equation:

$$\{\Delta P\} = \int [B]^T [D][\Delta C]\{\sigma\} dv \tag{20.3}$$

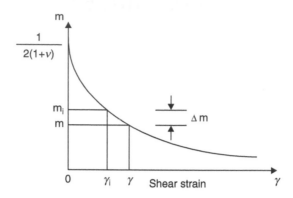

Figure 20.1 Relationship between increments of anisotropic parameter *m* and shear strain. (Sakurai, 1997b).

[K] is a stiffness matrix, [B] is a matrix of the relationship between strain and displacement and {σ} is stress in damaged zones where the anisotropic parameter *m* decreases with increase in shear strain. [D] and [ΔC] are given for a two-dimensional case in the following forms:

$$[D] = \frac{E}{1 - v - 2v^2} \begin{bmatrix} 1 - v & v & 0 \\ v & 1 - v & 0 \\ 0 & 0 & m(1 - v - 2v^2) \end{bmatrix} \tag{20.4}$$

$$[\Delta C] = \begin{bmatrix} 0 & 0 & 0 \\ 0 & 0 & 0 \\ 0 & 0 & \dfrac{\Delta m}{m \cdot m_i} \end{bmatrix} \tag{20.5}$$

The anisotropic parameters m_i and m, shown in Figure 20.1, can be back-calculated to minimise the error function given in Equation (2.3).

20.3   APPLICABILITY OF THE PROPOSED METHOD (FORWARD ANALYSIS)

In order to demonstrate the applicability of the proposed calculation procedure, a forward analysis is performed. A finite element mesh is shown in Figure 20.2, and the input data used in this forward analysis are shown in Table 20.1 (Sakurai & Hamada, 1997).

The displacement vectors are calculated and shown in Figure 20.3. In the example problem, the reduction of the anisotropic parameter *m* occurs in the shaded zone where sliding occurs, as shown in Figure 20.2.

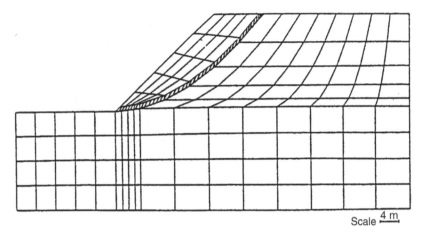

Figure 20.2 Finite element mesh.

Table 20.1 Input data for the forward analysis.

Young's modulus	E	100 MPa
Poisson's ratio	ν	0.3
Anisotropic damage parameter	m_i	0.10
	m	0.05

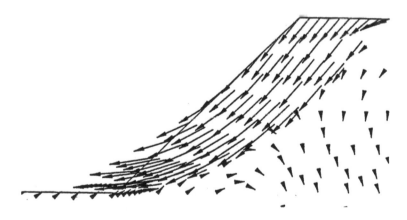

Figure 20.3 Displacement vectors calculated by numerical analysis.

In this example problem, the anisotropic parameter m and the sliding damaged zone are assumed, although they must be determined by back analysis (Sakurai and Hamada, 1996).

If the anisotropic parameter m is obtained from a back analysis of measured displacements during the excavation of slopes, we can immediately evaluate how much damage has already taken place in the slopes.

Figure 20.4 General view of slope.

20.4 CASE STUDY OF LANDSLIDE DUE TO HEAVY RAINFALL (BACK ANALYSIS)

A high-cut slope along a highway was monitored during excavation (see Figure 20.4). The displacement measurements were performed on the slope surface by using a total station. Large displacements were measured after heavy rainfall. At that time, there was no excavation being carried out. It is obvious that the displacements were caused by heavy rainfall, not by excavation. This deformational mechanism may be similar to landslides (Sakurai, 2001).

In order to stabilise the landslide, additional support measures were designed. For this purpose, the potential sliding plane must be first investigated by the proposed procedure given in Section 19.2. Once the potential sliding plane is determined, the finite element mesh can be easily made, resulting in back analysis being performed for determining the anisotropic parameter m and the increment Δm, shown in Figure 20.1, by using Equations (20.2)–(20.5). In this back calculation, matrix $[\Delta C]$ is a key to minimising the error function given in Equation (2.3). Once $[\Delta C]$ is determined, the displacements are easily calculated by Equations (20.2) and (20.3).

The back-calculated displacement vectors, shown in Figure 20.5, are compared with measured displacements. The result is shown in Figure 20.6. It is obvious from the figure that a good agreement between the two was obtained.

Once the anisotropic parameter m and Young's modulus E are back-calculated, the maximum shear strength, such as cohesion and internal friction angle along the potential sliding plane, can be determined by the back analysis procedure proposed in Section 21.2.

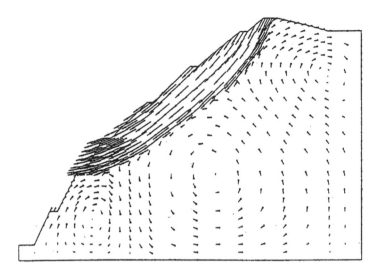

Figure 20.5 Displacement vectors back-calculated from the measured surface displacements. (Sakurai, 2001).

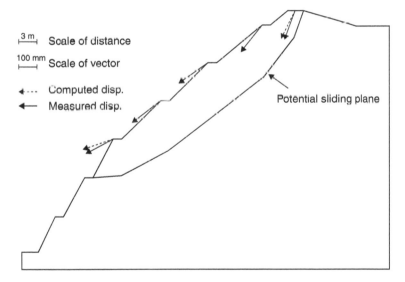

Figure 20.6 Comparison between back-calculated and measured displacements on the slope surface. (Sakurai, 2001).

Considering the back-calculated strength parameters, we can calculate the factor of safety along the potential sliding plane. If the calculated factor of safety is not large enough in comparison with its allowable value, then additional support measures, such as rock anchors and piles, are installed to secure the sufficient factor of safety. Since the location and shape of the potential sliding plane are back-calculated, the length of anchors is easily designed for stabilising the slope.

Chapter 21

Back analysis for determining the strength parameters

21.1 INTRODUCTION

In order to overcome the paradoxes of design and monitoring of slopes, monitoring results such as measured displacements should be linked with the strength parameters, such as cohesion and internal friction angle of geomaterials. To achieve this linkage, Sakurai and Nakayama (1999) proposed a back analysis procedure for determining the strength parameters from measured displacements. Once the strength parameters are determined, the factor of safety of slopes can be calculated (Sakurai et al., 2009) and the original design can be verified. If the back-calculated factor of safety is still sufficiently large, then the slopes must be adequately stable for the construction to proceed. However, if the factor of safety is less than that adopted at the design stage, additional support measures, such as drainage borings, earth anchors and piles, should be installed to stabilise the slopes. In addition, the design of the additional support measures is possible using the strength parameters determined by back analyses. The proposed back analysis procedure for determining the strength parameters of geomaterials from measured displacements is described in Section 21.2.

21.2 BACK ANALYSIS PROCEDURE

1   In the case of sliding, a potential slip plane can be back-calculated from the displacements measured on the ground surface (see Section 19.2). Once the potential slip plane is determined, a finite element mesh can be easily made by referring to it, as shown in Figure 18.13(a), (b) and (c). In the case of toppling, a finite element mesh is made by dividing the medium above the baseline in a number of layers, as shown in Figure 18.13(d). However, a question may arise about how to determine the angle of the baseline (the base angle α defined in Figure 18.6). To answer the question, it can be estimated considering the direction of displacement vectors measured on the slope surface, even though the accuracy of the estimation may not be a high quality. However, if we have some data of borehole inclinometers, then the base angle may be evaluated, resulting in that the accuracy of the estimation of the base angle may increase.

In any case, if we use a very small finite element mesh, there is no need to worry about the direction of the damaged zones (slip planes). The direction and size of the damaged zones can be back-calculated from the measured displacements on the ground surface.

2   Young's modulus E, the anisotropic parameter m and the angle θ between the local and global coordinate systems are back-calculated from the measured displacements, whereas Poisson's ratio v is assumed to be always constant. In the back analyses, the direct approach described in Section 2.3.3 is used in such a way that the error function defined by Equation (2.3) is minimised by changing the parameters E, m and θ. However, it should be noted that, in minimising the error function, the relationship between the anisotropic parameter and the maximum shear strain given in Equation (14.9) must be always satisfied during an iteration process.

Also, note that it is time consuming to back-calculate Young's modulus and the anisotropic parameter simultaneously. Thus, for an easy calculation, we can use Young's modulus on intact specimens determined by laboratory tests, because Young's modulus is insensitive for the factor of safety, as shown in a case study (see Section 22.2).

3   When Young's modulus and anisotropic parameter are obtained, shear modulus G can be determined by the following equation which is based on the definition of the anisotropic parameter:

$$G = mE \tag{21.1}$$

4   Once the shear modulus G is determined by Equation (21.1), the critical shear strain γ_0 can be obtained by Equation (21.2), which corresponds to the centre-line (average value) of the scattering data of the critical shear strains shown in Figure 21.1:

$$\gamma_0 = 4G^{-2/7} \tag{21.2}$$

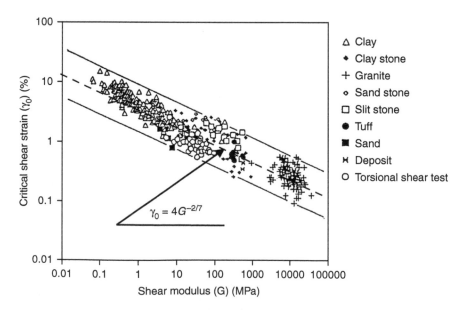

Figure 21.1 Criterion for assessing the stability of slopes.

The figure is based on Figure 7.3 and shows the relationship between the shear modulus G and the critical shear strain γ_0.

Since the data of the critical shear strains indicate large scattering, the centreline of the scattering data is proposed as the criterion for the factor of safety, FS = 1.0. The criterion corresponds to warning level II in assessing the stability of the tunnel (see Figure 7.5). As already described in Section 5.3, if the measured maximum shear strains remain less than warning level II, the tunnels are excavated without any problems; if the measured maximum shear strains become greater than warning level II, plastic zones may appear around the tunnels, but they are still stable because the plastic zones are usually supported by the surrounding elastic zones. However, in slopes when the maximum shear strains along the potential slip plane reach warning level II, failure may occur because there is no way to support the sliding block.

5 Maximum shear strength τ_c can then be calculated by the following equation, which is based on the definition of the critical shear strain:

$$\tau_c = G\gamma_0 \tag{21.3}$$

6 The internal friction angle ϕ of geomaterials can be determined by considering the relationship between the direction of damaged zones (slip planes) and the direction of principal stresses. Since a stress state in the damaged zones always satisfies the Mohr-Coulomb failure criterion, it is obvious that the direction of damaged zones is oriented in the direction of the angle $45° + \phi/2$ (where ϕ is the internal friction angle of geomaterials) from the direction of the minimum principal stress. Since the direction of local coordinate x'-axis is taken to be parallel to the direction of the damaged zones, it is possible to evaluate the internal friction angle provided that the angle θ between the local and global coordinates is known. In this context, the proposed back analysis method (see Section 19.2) is useful, as it can provide the angle θ, resulting in the internal friction angle being determined.

On the other hand, for toppling of slopes we need to identify the baseline, but the base angle α cannot be determined by a similar method to that proposed for sliding. As an alternative, therefore, the angle θ should be back-calculated from the measured displacements together with the anisotropic parameter m. Once the angle θ is determined, the internal friction angle ϕ can be determined in the same way as that for sliding, as described above.

Nevertheless, regardless of the deformational modes, sliding or toppling, the angle θ should be back-calculated from the measured displacements, resulting in the internal friction angle being determined.

7 Once the internal friction angle ϕ is determined, the cohesion c of geomaterials can be easily determined by the following equation:

$$c = \frac{1 - \sin\phi}{\cos\phi}\tau_c \tag{21.4}$$

8   The factor of safety can then be calculated using a conventional limit equilibrium method, such as the Fellenius method, as follows:

$$FS = \frac{\sum S_i \ell_i}{\sum \tau_i \ell_i} = \frac{\sum (c_i + \sigma_i \tan \phi_i) \ell_i}{\sum \tau_i \ell_i} \qquad (21.5)$$

where ℓ_i is the length of the sliding plane of the i-th slice, and σ_i and τ_i are the normal and shear stress on the sliding plane of the i-th slice, respectively.

Chapter 22

Application of back analysis for assessing the stability of slopes

22.1 CUT SLOPE

22.1.1 Introduction

Displacements are measured for assessing the stability of a cut slope that appeared in the construction of a newly developed golf course. The slope consists of highly weathered sandstone and slate, and the height and width of the slope are about 120 and 200 m, respectively. Global Positioning System (GPS) is used for measuring displacements. Two benchmarks are set about 600 m away from the slope and ten measuring points are set on the slope. Photo 22.1 shows the site and a GPS receiver and antenna on the tripod set at benchmark B. The GPS receivers used here are Trimble model 4000SST (8 channels, C/A code receiver) (Sakurai et al., 1992).

Figure 22.1 shows displacement vectors measured at the measuring points set on the slope. It is obvious from the figure that large displacements appear around measuring points 7–9.

Photo 22.1 GPS receiver and antenna, and the slope monitored by GPS.

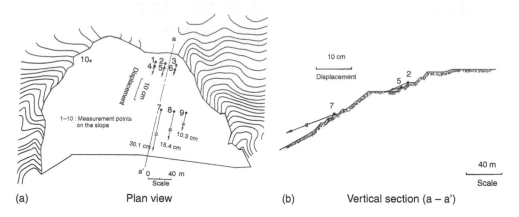

(a)　　　　　　　　　Plan view　　　　　　　(b)　　　　　Vertical section (a – a')

Figure 22.1　Displacement vectors on the slope measured by GPS.

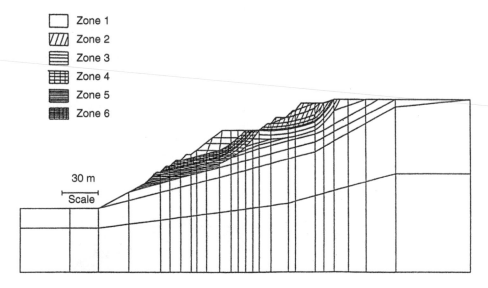

Figure 22.2　Finite element mesh.

22.1.2   Modelling and back analysis

Since rock masses of this site consist of jointed rock masses (joints are randomly oriented) of highly weathered sandstone and slate, the rock masses are modelled as a continuous body, so the constitutive Equation (14.19) is adopted in the back analyses. In the constitutive equation, the anisotropic parameter m plays a major role in expressing the non-linear mechanical behaviour of rocks and is back-calculated from the measured displacements to minimise the error function given in Equation (2.3).

Considering large displacements occurring at measuring point 7, the $a–a'$ line is chosen as a reference line whose vertical section is shown in Figure 22.2 together with

Table 22.1 Results of back analyses.

Young's modulus E	94 MPa
Unit weight γ (assumed)	19.6 kN/cm³
Poisson's ν (assumed)	0.30
Anisotropic parameter m	
m_1 (zone 1)	0.385 (isotropic)
m_2 (zone 2)	0.360 ($\alpha_2 = 45°$)
m_3 (zone 3)	0.100 ($\alpha_3 = 35°$)
m_4 (zone 4)	0.050 ($\alpha_4 = 35°$)
m_5 (zone 5)	0.012 ($\alpha_5 = 20°$)
m_6 (zone 6)	0.005 ($\alpha_6 = 20°$)
Shear modulus (sliding zone)	
G_4 (zone 4)	4.7 MPa
G_5 (zone 5)	1.1 MPa
Critical shear strain	
γ_{04} (zone 4)	1.0%
γ_{05} (zone 5)	1.6%
Internal friction angle	
ϕ (assumed)	25°
Cohesion	
c_4 (zone 4)	3.0×10^{-2} MPa
c_5 (zone 5)	1.1×10^{-2} MPa
Factor of safety FS	1.1

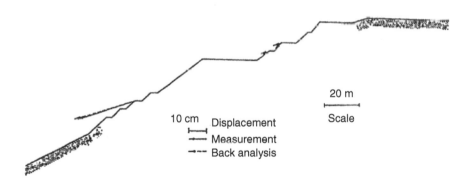

Figure 22.3 Comparison between calculated and measured displacements.

a finite element mesh. Back analyses are then performed under a two-dimensional plane strain condition for determining the strength parameters of rocks, following the back analysis procedure described in Section 21.2. The results of back analyses are summarised in Table 22.1.

Calculated displacements using the back-calculated mechanical parameters are compared with the measured values, as shown in Figure 22.3. It is obvious from the figure that there is a good agreement between the calculated and measured displacements. The displacement distribution and the maximum shear strain distribution of the slope determined by the back analysis are shown in Figures 22.4 and 22.5, respectively.

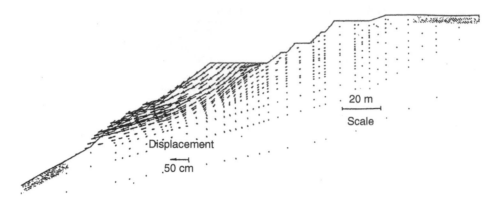

Figure 22.4 Displacements calculated by using back-analysed material constants.

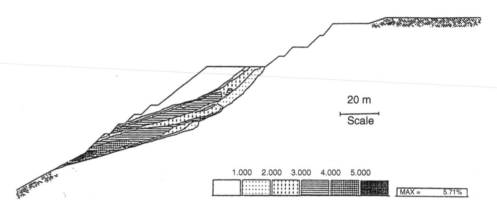

Figure 22.5 Maximum shear strain distribution.

22.1.3　Assessment of slope stability

The stability of slopes is assessed by the factor of safety defined by Equation (17.1). In order to assess the stability of slopes during constructions, displacement measurements are commonly performed, followed by back analysis of the measured displacements to determine strength parameters such as cohesion and internal friction angle of geo-materials. Once the strength parameters are determined, the factor of safety can be calculated by a conventional limit equilibrium method. The back analysis procedure for determining the factor of safety from measured displacements is described in Chapter 21.

In the concerned case study, a potential slip plane was presumed by considering the displacements measured by GPS, as shown in Figure 22.6, and the proposed back analysis was carried out to determine the factor of safety. All the results determined by back analysis are shown in Table 22.1. The value of the factor of safety is 1.1, which

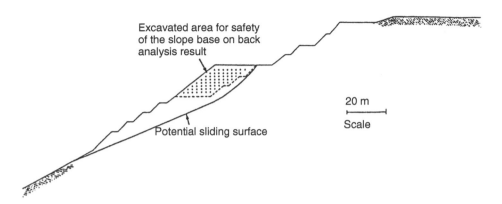

Figure 22.6 Potential sliding surface estimated by back analysis.

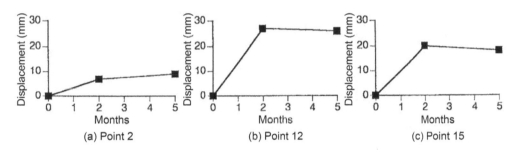

Figure 22.7 Displacements after changing the slope angle.

seems to be rather small, so for increasing the safety factor some part of geomaterials, as shown in the dotted zone in Figure 22.6, was excavated in such a way that the slope angle changed from 1:1 to 1:1.5 to stabilise the slopes.

In fact, GPS displacement measurements were continued for monitoring the slope stability after changing the slope angle. The results of measured displacements at the measuring points 2, 12 and 15 (12 and 15 are new measuring points; see Figure 22.6) are shown in Figure 22.7. As seen in the figure, the movement of displacements has stopped, resulting in the slope being completely stable.

22.2 SLOPE OF OPEN-PIT COAL MINE

22.2.1 Introduction

The proposed back analysis procedure was applied to assess the stability of an open-pit coal mine (height of the slope was measured as approximately 80 m). The displacements of the slope were measured by GPS, and the back analyses were carried out for determining strength parameters such as cohesion and internal friction angle of the geomaterials from measured displacements, resulting in the factor of safety being

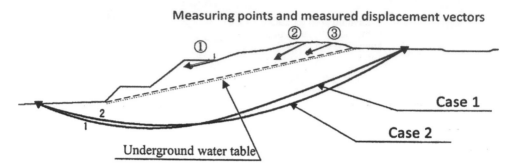

Figure 22.8 Displacement vectors measured by GPS together with the potential slip planes (Case I and Case 2).

calculated (see Section 21.2). The back-calculated factor of safety was approximately 1.0. In fact, the slope failed on the day after measuring the displacements that were used as data for the back analyses. This demonstrates that the proposed back analysis procedure can be applied to evaluating the stability of slopes with sufficient accuracy (Sakurai et al., 2009).

22.2.2 Cross section together with measuring points in the open-pit coal mine

The cross section of the slope with the GPS measuring points is shown in Figure 22.8. In this case study, we used the two potential slip planes (Case 1 and Case 2) as shown in the figure, which were predicted by the proposed procedure described in Section 19.2. The two slip planes were obtained by a second-order polynomial equation changing its parameters. The displacement vectors measured by GPS at the ground surface are also shown in the figure.

22.2.3 Input data for the back analysis

The following input data were used:

1 Failure criterion: Mohr–Coulomb type.
2 Young's modulus: Three different values, i.e. $E = 28,000$, $56,000$ and $112,000 \, \text{kN/m}^2$, were used because no information was available.
3 Poisson's ratio: $\nu = 0.3$.
4 Underground water table: It is shown in Figure 22.8, and was predicted by field observations on the ground surface.
5 Thickness of sliding plane: $t = 1$ and $2 \, \text{m}$ (assumed).
6 Displacement data: The data (GPS measurements) were those obtained one day before the failure occurred. The horizontal and vertical components of displacement vectors were measured at three measuring points – ①, ② and ③ – on the ground surface, as shown in Figure 22.8.

22.2.4 Back analysis procedure and the results

Back analyses were performed using the proposed procedure to obtain the strength parameters of geomaterials from the displacements measured by GPS on the ground surface, resulting in the factor of safety of slope being determined, as described in Section 21.2.

1 First, a potential slip plane was back-calculated from the measured displacements by the proposed method, as described in Section 19.2.
2 Back analyses were then carried out, where Young's modulus and Poisson's ratio are supposed to be one of the assumed values, and an attempt is made to minimize the error function shown in Equation (22.1) by changing anisotropic parameter m until obtaining its optimal value.

$$\Delta u^2 = \sum_{i=1}^{M} (u_i^c - u_i^m)^2 \to \min. \tag{22.1}$$

where u_i^c: Computed displacements, u_i^m: Measured displacements, M: Number of measurement data.
3 Once the optimal value of the anisotropic parameter m is obtained, the strength parameters (c and ϕ) of the geomaterials can be determined by the proposed back analysis procedure, described in Section 21.2.
4 The factor of safety, Fs, can then be calculated by a conventional limit equilibrium method, shown in Equation (21.5).

22.2.5 Results of the back analysis

Using the input data given in Section 22.2.3, the back analyses were carried out for Case 1 and Case 2, resulting in the factor of safety being calculated. In the back analyses, the optimal value of the anisotropic parameter m is determined to minimise the error function defined by Equation (22.1). The back analysis results reveal that the error function for Case 1 yields a smaller value than that for Case 2. This means that the slip plane for Case 1 is more realistic than that for Case 2 in this open-pit mine.

The back analysis results for Case 1 are shown in Table 22.2, where Young's modulus is assumed to be three different values. As a reference, the horizontal and vertical components of back-calculated displacements, together with the measured displacements at the three measuring points, are also shown in Table 22.2. Regarding the thickness of the slip plane, the back analysis results shown in Table 22.2 indicate that $t = 1$ m seems to be preferable, because the factor of safety for the case of $t = 2$ m is slightly larger than that for $t = 1$ m. Since the slope failed on the day after data acquisition, the value of the factor of safety should be very close to 1.0 just before the failure occurred. Considering this fact, the thickness of the slip plane may be nearly $t = 1$ m. This is valuable information for assessing the stability of slope of this open-pit mine in the future.

One of the back analysis results for Case 1 is shown in Figure 22.9, where the back-calculated displacement vectors are compared with those of the measured displacements.

Table 22.2 Back analysis results for displacements at the measuring points (Case I).

Thickness of sliding plane (m)	E (kN/m²)	m	FS	Δu^2 (cm²)	Displacements at measuring points (cm)					
					① δ_x 168.0	① δ_z 46.0	② δ_x 183.3	② δ_z 96.3	③ δ_x 164.4	③ δ_z 69.8
1.0	28,000	0.0045	1.073	83	170.0	53.0	172.1	82.8	173.2	77.4
	56,000	0.0023	1.004	95	169.5	52.7	173.3	79.9	174.1	77.4
	112,000	0.0012	0.959	107	169.2	52.3	174.5	77.8	175.2	77.7
2.0	28,000	0.0085	1.218	81	170.9	53.5	172.2	85.6	172.8	80.3
	56,000	0.0045	1.120	95	167.3	52.0	171.4	79.6	172.3	77.1
	112,000	0.0023	1.051	105	169.5	52.5	174.4	78.5	175.1	78.3

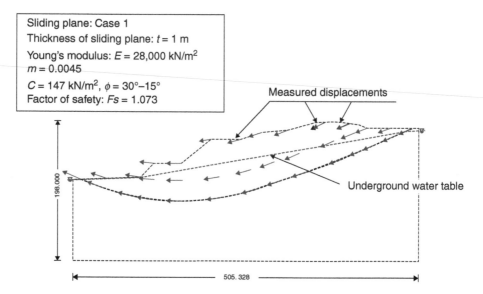

Sliding plane: Case 1
Thickness of sliding plane: $t = 1$ m
Young's modulus: $E = 28,000$ kN/m²
$m = 0.0045$
$C = 147$ kN/m², $\phi = 30°–15°$
Factor of safety: $Fs = 1.073$

Measured displacements

Underground water table

505. 328

Figure 22.9   Comparison of computed and measured displacement vectors.

22.2.6   No-tension analysis

According to field observations at the mine site, some vertical open cracks appeared at the end of the potential slip plane even before the slope failure. Therefore, the vertical open cracks should be taken into account in the back analyses. For this purpose, no-tension analyses were carried out considering the vertical open crack with the depth of 8 m (Zienkiewicz et al., 1968). One of the back analysis results for the no-tension analysis is shown in Figure 22.10, which shows that the factor of safety is 0.993. It is obvious that a vertical open crack with a depth of 8 m slightly reduces the factor of safety.

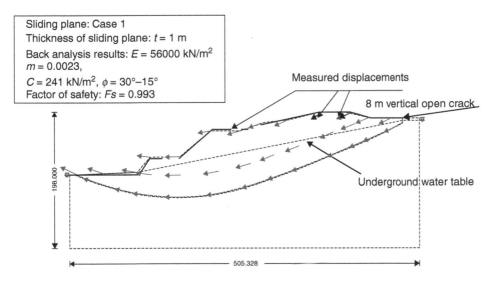

Sliding plane: Case 1
Thickness of sliding plane: $t = 1$ m
Back analysis results: $E = 56000$ kN/m^2
$m = 0.0023$,
$C = 241$ kN/m^2, $\phi = 30°-15°$
Factor of safety: $Fs = 0.993$

Measured displacements

8 m vertical open crack

198.000

Underground water table

505.328

Figure 22.10 No-tension analysis: Comparison between the computed and measured displacement vectors.

22.2.7 Discussions on the back analysis results

The anisotropic parameter m is back-calculated from measured displacements under the assumption of Young's modulus E. In this case study, three different values of Young's modulus were assumed. It is obvious that the different values of anisotropic parameters are obtained for the different values of Young's modulus. However, it is of interest to see in Table 22.2 that if a large value of Young's modulus is assumed, then a small value of the anisotropic parameter is back-calculated, and vice versa, resulting in the shear modulus derived as $G = mE$ (from the definition of m) becoming more or less identical, no matter what values of Young's modulus are assumed. It is a great advantage that any value of Young's modulus can be used for determining the factor of safety from measured displacements. This means that Young's modulus of geomaterials determined by laboratory tests can be used directly in the engineering practice without taking into account the size effect of specimens.

It must be noted, however, that the real value of Young's modulus of geomaterials cannot be determined by the proposed back analysis procedure, whereas strength parameters such as cohesion and internal friction angle of geomaterials determined by the proposed back analyses are the real values for the geomaterials. As a reference, if it is necessary to determine the real value of Young's modulus of geomaterials from measured displacements, we should use some other back analysis procedure; for example, the back analysis procedure based on the normalised initial stresses can be used, as described in Section 8.2.

It is also of interest to see in Table 22.2 that the factor of safety of the slope is approximately 1.0, no matter what values of Young's modulus are assumed. This demonstrates that the slope must be in a very critical state in terms of its stability. In fact, the slope failed on the day after taking the displacement data that were used for the

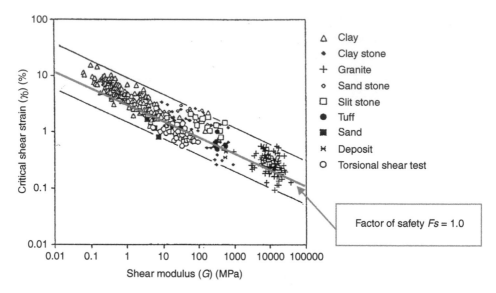

Figure 22.11   Factor of safety is 1.0 for slopes in relation to the shear modulus and the maximum shear strain.

back analyses. This implies that the proposed back analysis procedure can determine the factor of safety with sufficient accuracy only from surface displacements measured by GPS.

In this case study, the centreline of the scattering data of the critical shear strain, as seen in Figure 21.1, was used as a criterion for assessing the stability of slopes. Consequently, if the maximum shear strain reaches the centreline of the scattering data, the slopes may fail along the slip plane. In other words, the centreline corresponds to the critical state of slopes, i.e. the factor of safety becomes 1.0, as shown in Figure 22.11. As already described in Section 7.2, the centreline of the scattering data may be used as a criterion for assessing the stability of tunnels. Figures 5.5 and 5.6 demonstrate that the centreline is the boundary of the troublesome and untroubled tunnels during excavations.

It is noted that the centreline (average value) of the scattering data of critical shear strain must be an important design criterion for assessing not only tunnels but also slopes, resulting in the stability assessment of both tunnels and slopes being consistent with each other.

Chapter 23

Monitoring of slope stability using GPS in geotechnical engineering

23.1 INTRODUCTION

The ideal monitoring system for projects in rock and geotechnical engineering should be able to continuously and automatically monitor the behaviour of an extensive area in real time and with high accuracy. In addition, the costs should be low and the handling should be easy. Displacement monitoring using GPS satisfies the above requirements.

GPS began to be used for displacement monitoring in the mid-1980s in civil and mining engineering as well as other related fields. Since then, practical applications have been performed and guidelines have been published for displacement monitoring. The International Society for Rock Mechanics (ISRM) has also approved the suggested method for monitoring rock displacements using GPS (Shimizu et al., 2014).

In this section, the procedure for monitoring displacements using GPS is outlined, together with error-correction methods based on *ISRM Suggested Methods* (Shimizu et al., 2014).

23.2 DISPLACEMENT MONITORING USING GPS

GPS is a satellite-based positioning system that was developed by the U.S. Department of Defense (Hoffman-Wellenhof et al., 2001; Misra & Enge, 2006). Recently, similar systems are being operated by Russia, the European Union, China, and Japan. Those including GPS are collectively called the Global Navigation Satellite System (GNSS).

GPS provides the three-dimensional relative coordinates in latitude, longitude and height between two points with the accuracy of millimetres to centimetres. By continuously observing the coordinates of measurement points, the displacements are obtained as changes in the coordinates. The standard deviation of the measurements can be just a few millimetres when the baseline length is less than 1 km, and the installation and data corrections are conducted carefully.

23.2.1 Monitoring procedure

The general procedure for measuring displacements by GPS is as follows:

1   Set up the receiver(s) at the measurement point(s) and the reference point (Figure 23.1).
2   Measure the carrier phase of the signal and receive the navigation data at each measurement point and reference point from the same satellites simultaneously. The measured data are saved automatically in the memory of the receivers.

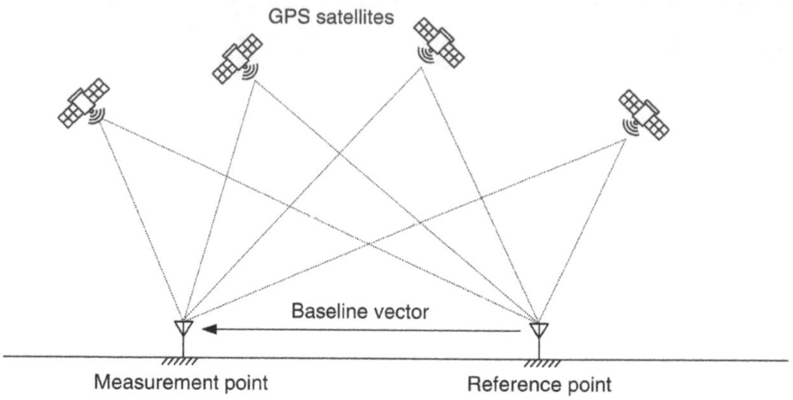

Figure 23.1 **GPS measurements.**

3   Download the data from all the receivers to a computer.
4   Collect the downloaded data of the reference point and the measurement point(s), and conduct a baseline analysis to obtain the relative coordinates of the measurement point(s) from the reference point.
5   Repeat Steps 3 and 4 for the succeeding observation session.
6   Calculate the changes in the coordinates at each measurement point between the two sessions, and then obtain the displacements. Software for data analysis is provided by the manufacturer of the receiver.

When standard receivers for surveys are employed to measure displacements, the users are usually required to conduct the above procedure manually. Then, it is inconvenient and ineffective for continuous monitoring. In order to overcome this troublesome process, an automatic monitoring system was designed and developed as illustrated in Figure 23.2 (Iwasaki et al., 2003; Masunari et al., 2003; Shimizu et al., 2011), and applied for monitoring displacements in various projects.

Sensors, composed of an antenna and a terminal box, are set on measurement points and a reference point. They are connected to a control box into which a computer, a data memory and a network device are installed. The data emitted from the satellites are received at the sensors and then transferred to the control box through cables. The server computer, which is located in an office away from the measurement area, automatically controls the entire system to acquire and then analyse the data from the control box. Then, the three-dimensional displacements at all the monitoring points are obtained. The monitoring results are provided to users on the web through the Internet in real time. The users only need to access the home page to see the monitoring results.

23.2.2   Improvements in accuracy: Error corrections

Standard deviations in the conventional GPS survey are 5 mm in the horizontal direction (width) and 10 mm in the vertical direction (height) when the baseline length is less than 1 km. Precise monitoring for practical use, however, often requires higher accuracy than this. The most important issue in the practical use of GPS is how to improve measurement accuracy.

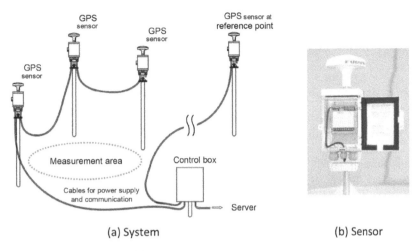

(a) System   (b) Sensor

Figure 23.2 GPS displacement monitoring system (Shimizu et al., 2011).

Measurement results generally include random errors (noise) and bias errors. Random errors arise from random fluctuations in the measurements. On the other hand, typical bias errors in GPS monitoring are tropospheric delays and other signal disturbances due to obstructions above the antennas and multipath effects. Since both random and bias errors affect the monitoring quality, it is recommended that appropriate error-correction methods be applied to reduce such errors.

Generally, the measurement value y is composed of exact value u, bias errors ε_p and ε_T, and random error ε_R, as follows:

$$y = u + \varepsilon_p + \varepsilon_T + \varepsilon_R \tag{23.1}$$

where ε_p is an error due to a signal disturbance caused by obstructions above the antennas and multipath effects, ε_T is an error due to tropospheric delays and ε_R is a random error due to receiver noise. Tropospheric delays occur when signals from the satellite reflect in the atmosphere due to meteorological conditions (Hoffman-Wellenhof et al., 2001; Misra & Enge, 2006).

Both bias errors ε_p and ε_T can be reduced and corrected by methods described in *ISRM Suggested Methods* (Shimizu et al., 2014). After removing the bias errors from the measurement value, an appropriate statistical model is applied to estimate the exact displacement.

23.3 PRACTICAL APPLICATION OF GPS DISPLACEMENT MONITORING

The GPS displacement monitoring system (see Figure 23.2) has been applied to monitor the displacements of an unstable steep slope along a national road in Japan. Displacement monitoring was conducted by borehole inclinometers and surface extensometers.

Some of the instruments, however, occasionally did not work well due to large deformations, and it was difficult to perform the monitoring continuously. In order to overcome such trouble, the GPS monitoring system has been applied for continuous monitoring (Furuyama et al., 2014).

23.3.1 Monitoring site: Unstable steep slope

Figure 23.3 presents photographs of the slope and the monitoring area. The slope is composed of mainly tuff and sandstone, and its bedrock is granite. Two antennas were set at the top of the slope to monitor displacements, and another antenna was set at a fixed point in a stable area as a reference point, denoted by K-1, beneath the slope (see Figure 23.4). The monitoring points, denoted as G-1 and G-2, were set on the left and right sides of the slope, respectively.

(a) Slope and monitoring area (b) GPS Sensors at G-1 and G-2

Figure 23.3 Monitoring site and slope beside road (Furuyama et al., 2014).

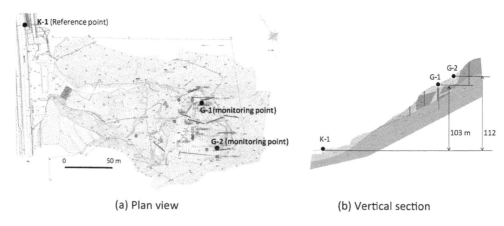

(a) Plan view (b) Vertical section

Figure 23.4 Monitoring points and reference point (Furuyama et al., 2014).

Three-dimensional displacements were continuously measured every hour using the static method of GPS. The distance between monitoring point G-1 and reference point K-1 was 221 m, whereas that between monitoring point G-2 and the reference point was 258 m. The differences in height between the two points and the reference point were 103 and 112 m, respectively.

23.3.2 Effects of error corrections

Since both differences in height between reference point K-1 and monitoring points G-1 and G-2 were more than 100 m, the error due to the tropospheric delays of the signal wave from the satellite could not be ignored. In addition, since the trees around the monitoring points will grow and become obstructive above the antennas, they will cause a disturbance to the signals and a decline in the measurement accuracy. In order to correct these types of errors, the error-correction methods indicated in Section 23.2.2 were applied (Shimizu et al., 2014).

Figure 23.5(a) shows the original monitoring results obtained from a baseline analysis using the observed wave phase. Although the slope did not move at all during this period, the results were rather wavy and scattered.

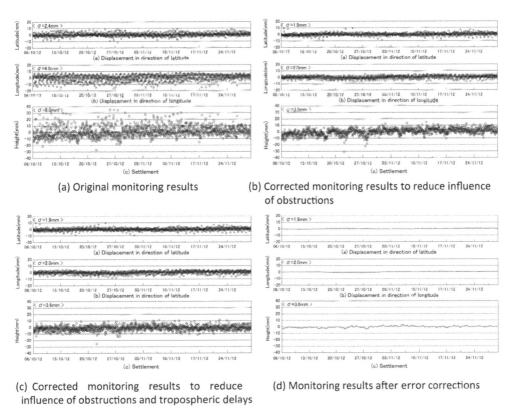

(a) Original monitoring results

(b) Corrected monitoring results to reduce influence of obstructions

(c) Corrected monitoring results to reduce influence of obstructions and tropospheric delays

(d) Monitoring results after error corrections

Figure 23.5 Effects of error corrections for GPS displacement monitoring results (Furuyama et al., 2014).

As the first step of correction, the signals obtained from the satellites located behind the obstructions above the GPS antenna were not used when analysing the data. Figure 23.5(b) presents the corrected results. Comparing the original and the corrected results (Figures 23.5(a) and 23.5(b)), the scattering that appeared in the original results significantly improved; i.e. the standard deviations of the displacements in the direction of latitude, longitude and height improved from 2.4, 4.6 and 8.0 mm to 1.9, 2.0 and 3.5 mm, respectively. Some periodic behaviour, however, still remains.

As the second step of correction, an appropriate model was employed to correct the tropospheric delays. The modified Hopfield model, which is used for estimating tropospheric delay (Hoffman-Wellenhof et al., 2001; Misra & Enge, 2006), was adopted. Figure 23.5(c) shows the results after correcting the obstructions and tropospheric delays. Comparing the original and the corrected results (Figures 23.5(a) and 23.5(c)), the scattering and the periodic behaviour appearing in the original results were clearly improved.

Finally, as the third step of correction to reduce random errors, the 'trend model' was applied to estimate the exact displacement from the monitoring results (Shimizu et al., 2014). The solid lines in Figure 23.5(d) show the final results. Although some small scattering still exists, it is found from these results that no displacement occurred during this period.

Comparing the original and the corrected monitoring results (Figures 23.5(a) and 23.5(d)), it can be seen that the original results significantly improved through the above error-correction procedure.

23.3.3 Monitoring results

The displacement monitoring results at monitoring point G-1 are shown with an hourly amount of rainfall in Figure 23.6. Monitoring was started in the middle of March 2013. The three-dimensional displacements in the direction of latitude, longitude and height were continuously measured every hour.

The standard deviations of the measurements at G-1 and G-2 are shown in Table 23.1. It is shown that the GPS monitoring system and the correction method used here have been successful. The monitoring results have been obtained continuously and smoothly without missing any results.

Figure 23.7(a) shows the displacement vectors in the plan view of the slope, whereas Figures 23.7(b) and 7(c) show the vertical sections including G-1 and G-2, respectively. Both directions of vectors for G-1 and G-2 almost coincided with the steepest direction of the slope in the plan view. The direction of the displacement at G-1 was towards the front of the slope in the vertical section until the middle of August. After heavy rainfall at the end of August and early September, the direction changed to be parallel to the slip plane of the slope, and it was towards the front of the slope again after the last large displacement of 23 October. On the other hand, the direction of the displacement at G-2 was parallel to the slip plane at the top of the slope in the vertical section.

The GPS displacement system (Figure 23.2) has been applied to monitor landslides, cut slopes, quarries, dams, tunnel entrances, viaducts, retaining walls, railway tracks etc. (Shimizu and Nakashima, 2017).

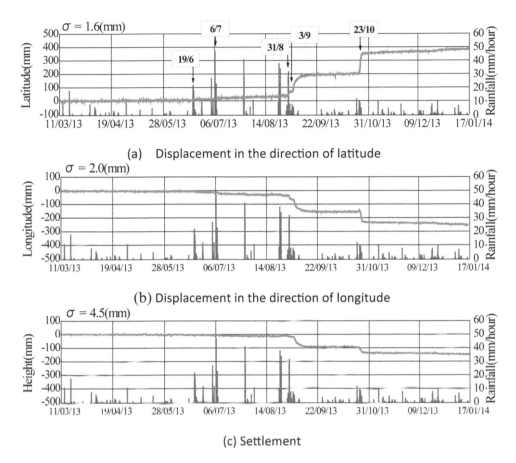

(a)   Displacement in the direction of latitude

(b) Displacement in the direction of longitude

(c) Settlement

Figure 23.6 Displacement monitoring results at point G-1 (Furuyama et al., 2014).

Table 23.1 Standard deviations in measurement.

Measurement point	Latitude (mm)	Longitude (mm)	Height (mm)
G-1	1.6	2.0	4.5
G-2	1.9	1.6	3.6

(Monitoring duration: March 2013 to January 2014)

23.4   BACK ANALYSIS IN GPS DISPLACEMENT MONITORING

Various types of instruments, such as extensometers and inclinometers, are available for measuring displacements in geotechnical engineering practice. However, many of them are applied to measure displacements only in a short distance, i.e. less than

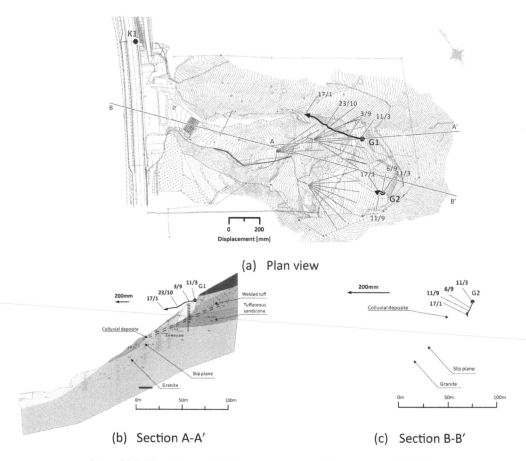

(a) Plan view

(b) Section A-A'

(c) Section B-B'

Figure 23.7 Transitions of displacement vectors (Furuyama et al., 2014).

100 m, so they may be inadequate to monitor the stability of large slopes and/or ground movements over an extensive area. In addition, many of them can measure only one-dimensional displacements in the direction of a measurement line or perpendicular to it. To overcome this shortcoming, total stations, electronic distance meters, levels etc. can be used for measuring three-dimensional displacements in an extensive area. However, they generally require a lot of time, labour and financial resources. In addition, their accuracy is not good enough to perform precise measurements as a geotechnical monitoring technique.

On the other hand, recent technologies of radar and laser have been developed remarkably, such that satellite, aircraft and ground-based Interferometric Synthetic Aperture Radar (InSAR), and laser profiler are often applied for measuring ground surface deformations. They are useful for finding large deformation areas and mapping unstable landslides in a large area. However, all these techniques are also those for measuring one-dimensional displacements between instruments and targets. In other words, the displacements measured by these technologies are not 'vector' but 'scalar' quantities.

According to the definition of 'monitoring', measurements must be continuously carried out. In addition, an 'alarm' should be issued on the basis of measurement results. In order to issue the alarm for warning about the critical state of slopes before their failure, the measurement data should be properly interpreted. One of the conventional stability assessments of slopes from the measured displacements is based on an engineer's judgement. For example, if the measured displacements start accelerated, then certain support measures are installed to stabilise them. According to the engineer's judgement, however, the installation of support measures often becomes too late to stabilise the slopes, resulting in aggressive support measures being necessary. Considering these situations of monitoring, it should be noted that the measurement data must be quantitatively assessed during the course of monitoring.

For quantitative assessments of the stability of slopes, the factor of safety is determined by the back analysis of measured displacements. However, for back analyses, the measured displacements should be three-dimensional displacement vectors. If the measured displacements are one-dimensional displacements (scalar quantities), such as the results of InSAR and laser profiler, the factor of safety cannot be calculated. In order to issue the alarm for the possible failure of slopes from measured displacements, three-dimensional measured displacements (vector quantities) are essential. If the three-dimensional displacement measurement results are available, the quantitative assessment of slope stability can be performed by back analyses, resulting in the alarm being issued.

To measure three-dimensional displacements, GPS is a powerful tool that can perform continuous and automatic measurements to obtain three-dimensional displacements over an extensive area including large slopes with high accuracy. GPS monitoring results can be used as input data for back analyses to provide the factor of safety and quantitatively assess the stability of slopes, resulting in GPS displacement measurements providing an alarm for warning about the failure of slopes.

References

Adachi, T., Serata, S. & Sakurai, S. (1969) Determination of underground stress field based on inelastic properties of rocks. *Rock Mechanics—Theory and Practice (Proc. 11th Symposium on Rock Mechanics, 16–19 June, The University of California, Berkeley, California)*, Society of Mining Engineers, Berkeley, California, 293–328.

Akayuli, C.F.A. & Sakurai, S. (1998) Numerical analyses of the failure pattern of two parallel shallow tunnels excavated in granular materials. *Memoirs of No. 40-B, November "Special Issue of The 1995 Hanshin-Awaji Great Earthquake", Construction Engineering Research Institute Foundation, Kobe University*, Kobe, Japan, 153–164.

Astrom, K.J. & Eykhoff, P. (1971) System identification: A survey. *Automatica*, 7, 123–162.

Cividini, A., Jurina, L. & Gioda, G. (1981) Some aspects of characterization problems in geomechanics. *Int. J. Rock Mechanics and Mining Sciences*, 18, 487–503.

Cividini, A., Maier, G. & Nappi, A. (1983) Parameter estimation of a static geotechnical model using a Bayes' approach. *Int. J. Rock Mechanics and Mining Sciences*, 30(3), 215–226.

Cividini, A., Gioda, G. & Barla, G. (1985) Calibration of a rheological material model on the basis of field measurements. In T. Kawamoto & Y. Ichikawa (Eds.). *Proc. 5th International Conference on Numerical Methods in Geomechanics*, 1–5 April, Nagoya, Japan, Rotterdam/Boston, A.A. Balkema, 1621–1628.

Cividini, A. & Gioda, G. (2003) Back analysis of geotechnical problems. In J.W. Bull (Ed.). *Numerical analysis and modeling in geomechanics*, Chapter 6, (pp. 165–196). London: Spon Press (Taylor & Francis).

Contini, A., Cividini, A. & Gioda, G. (2007) Numerical evaluation of the surface displacements due to soil grouting and to tunnel excavation. *International Journal of Geomechanics* (ASCE), 7(3), 217–226.

Cundall, P.A. (1971) A computer model for simulating progressive large-scale movements in blocky rock systems. *Proc. Sympo. Int. Soc. Rock Mech.*, Nancy, ISRM, II, 2–8.

Cundall, P.A. (1977) Computer interactive graphics and the distinct element method, *in Rock Engineering for Foundations and Slopes (Proc. the ASCE Specialty Conference, University of Colorado, 1976)*, ASCE, New York, Vol. 2: 193–199.

Deere, D.C. (1968) Geological considerations. In Staff, K.G. and Zienkiewicz, O.C. (Eds.), *Rock Mechanics in Engineering Practice*, New York, Toby Wiley & Sons, 1–20.

Deeswasmongkol, N. & Sakurai, S. (1985) Study on rock slope protection of toppling failure by physical modelings. *Proc. 26th U.S. Symposium on Rock Mechanics, Vol. 1, 26–28 June, Rapid city, South Dakota, USA*. Rotterdam, A.A. Balkema, 11–18.

Deeswasmongkol, N. & Sakurai, S. (1986) Some experiments on toppling failure by physical modelings. *Mem. Grad. School Sci. & Technol. Kobe Univ.*, 4-A, 9–20.

Drucker, D. C. & Prager, W. (1952) Soil mechanics and plastic analysis for limit design. *Quarterly of Applied Mathematics*, 10(2), 157–165.

Fairhurst, C. & Lin, D. (1985) Fuzzy methodology in tunnel support design. *Proc. 26th US Symposium on Rock Mechanics, Vol. 1, 26–28 June, Rapid city, South Dakota, USA*. Rotterdam, A.A. Balkema, 269–278.

Feng, X.-T., Zhang, Z. & Sheng, Q. (2000) Estimating mechanical rock mass parameters relating to the Three Gorge Project permanent shiplock using as intelligent displacement back analysis method. *Int. J. Rock Mechanics & Mining Sciences*, 37, 1039–1054.

Feng, X.-T. & Yang, C. (2004) Coupling recognition of the structure and parameters of nonlinear constitutive material models using hybrid evolutionary algorithms. *Int. J. Numerical Methods in Engineering*, 59, 1227–1250.

Feng, X.-T., Zhao, H. & Li, S. (2004) A new displacement back analysis to identify mechanical geo-material parameters based on hybrid intelligent methodology. *Int. J. Numerical and Analytical Methods in Geomechanics*, 28, 1141–1165.

Feng, X.-T., Chen, B.-R., Yang, C., Zhou, H. & Ding, X. (2006) Identification of visco-elastic models for rocks using genetic programming coupled with the modified particle swarm optimization algorithm. *Int. J. Rock Mechanics & Mining Sciences*, 43, 789–801.

Feng, X.-T. & Hudson, J. A. (2010) Specifying the information required for rock mechanics modeling and rock engineering design. *Int. J. Rock Mechanics & Mining Sciences*, 47, 179–194.

Fletcher, R. & Reeves, C.M. (1964) Function minimization by conjugate gradient. *The Computer Journal*, 7(2): 149–154.

Furuyama, Y., Nakashima, S. & Shimizu, N. (2014) Displacement monitoring using GPS for assessing stability of unstable steep slope by means of ISRM suggested method. *Proceedings of the 2014 ISRM International Symposium – 8th Asian Rock Mechanics Symposium (ARMS8), 14–16 October, Sapporo, Japan*, 1897–1904.

Gioda, G. (1980) Indirect identification of the average elastic characteristics of rock masses. *Proc. Int. Conf. on Structural Foundations on Rock, 7–9 May, Sydney*, Rotterdam, A.A. Balkema, 65–73.

Gioda, G. & Maier, G. (1980) Direct search solution of an inverse problem in elastoplasticity: Identification of cohesion, friction angle and *in situ* stress by pressure tunnel tests. *International Journal for Numerical Methods in Engineering*, 15, 1823–1848.

Gioda, G. & Jurina, L. (1981) Identification of earth pressure on tunnel liners. *Proc. 10th Int. Conference on Soil Mechanics and Foundation Engineering, 15–19 June, Stockholm*, Rotterdam, A.A. Balkema, 301–304.

Gioda, G. & Sakurai, S. (1987) Back analysis procedures for the interpretation of field measurements in geomechanics. *Int. J. Numerical and analytical methods in Geomechanics*, 11, 555–583.

Goodman, R.E. & Bray, J.W. (1976) Toppling of rock slopes. *Proc. Specialty Conf. on Rock Engineering for Foundations and Slopes, 15–18 August, Boulder, Colorado, USA*, 201–234.

Goodman, R.E. & Kieffer, D.S. (2000) Behavior of rock in slopes, *Journal of Geotechnical and Geoenvironmental Engineering*, 126(8): 675–684.

Goodman, R. E., Taylor, R. & Brekke, T. (1968) A model for the mechanics of jointed rock. *Journal of the Soil Mechanics and Foundations Division*, ASCE, 94(SM3), 637–659.

Hansmire, W.H. & Cording, E.J. (1985) Soil tunnel test section: Case history summary. *Journal of Geotechnical Engineering*, ASCE, 111(11), 1301–1320.

Hart, G.C. & Yao, J.T.P. (1977) System identification in structural dynamics. *J. Eng. Mech. Div.*, ASCE, 103 (EM6), 1089–1104.

Hoek, E. (1998) Tunnel support in weak rock. *Keynote address, Symposium of Sedimentary Rock Engineering, November 20–22, Taipei, Taiwan*.

Hoffman-Wellenhof, B., Lichtengger, H. & Collins, J. (2001) *GPS – Theory and Practice. 5th revised edition*. Vienna: Springer.

Inada, Y. & Kokudo, Y. (1992) Effect of high and low temperature on failure characteristics of rock under compression. *J. Soc. Mat. Sci., Japan*, 41(463), 410–416. (in Japanese)

Iwano, M., Otsuka, I., Taki, H., Harada, H. & Sakurai, S. (2010) Construction of underground railways station beneath unsound buildings in densely populated urban area, Istanbul, Turkey. *ITA-AITES World Tunnel Congress 2010 entitled "Tunnel Vision Towards 2020" Session 5C Tunnelling under Sensitive Structures, 14–20 May, Vancouver, Canada.*

Iwasaki, T., Takechi, K., Takeishi, A., Masunari, T., Takechi, Y. & Shimizu, N. (2003) Web-based displacement monitoring system using GPS for the maintenance of roadside slopes. F. Myrvoll (ed.), *Proceedings of the 6th International Symposium on Field Measurements in Geomechanics, FMGM03, 15–18 September, Oslo, Norway*, Leiden, CRC Press/Balkema, 137–143.

Kalman, R.E. (1960) A new approach to linear filtering and prediction problems. *Trans. A.S.M.E., J. Basic Eng.*, 82, 35–45.

Kavanagh, K. & Clough, R. (1971) Finite element application in the characterization of elastic solids. *Int. J. Solids Struct.*, 7, 11–23.

Kavanagh, K. (1973) Experiment versus analysis: Computational techniques for the description of static material response. *Int. J. Num. Meth. Eng.*, 5, 503–515.

Kawai, T. (1980) Some considerations on the finite element method. *Int. J. Numerical Methods in Engineering*, 16, 81–120.

Khamesi, H., Torabi, S.R., Mizaei-Nasirabad, H. & Ghadiri, Z. (2015) Improving the performance of intelligent back analysis for tunneling using optimized fuzzy systems: Case study of the Karaj Subway Line 2 in Iran. *J. Comput. Civ. Eng.*, ASCE, 29(6), 05014010-1.

Kirsten, H.A.D. (1976) Determination rock mass elastic moduli by back analysis of deformation measurements. In Z.T. Bieniawski (ed.), *Proc. of Symposium on Exploration for Rock Engineering, 1–5 November, Johannesburg, South Africa*, 165–172. Cape Town: A.A. Balkema.

Kohmura, Y. (2012) A study on critical strain of rocks (in Japanese). *Proc. of JSCE (Geotechnical Engineering), Japan Society of Civil Engineers*, 68(3), 526–534.

Kovari, K., Amstad, Ch. & Fritz, P. (1977) Integrated measuring technique for rock pressure determination. *Proc. Int. Symposium on Field Measurements in Rock Mechanics, 4–6 April, Zurich, Vol. 1*, 289–316.

Kovari, K. & Amstad, Ch. (1984) Fundamentals of deformation measurements. *Proc. Int. Symposium on Field Measurements in Geomechanics, Vol. 1, 5–8 September, Zurich, Switzerland*, 219–239.

Masunari, T., Tanaka, K., Okubo, N, Oikawa, H., Takechi, K., Iwasaki, T. & Shimizu, N. (2003) GPS-based continuous displacement monitoring system. *Proceedings of International Symposium on Field Measurements in Geomechanics, FMGM03, 15–18 Sept., Oslo, Norway*, 537–543.

Misra, P. & Enge, P. (2006) *Global Positioning System – signals, measurements, and performance. 2nd ed.* Lincoln, MA, Ganga-Jamuna Press.

Murakami, A. & Hasegawa, T. (1985) Observational prediction of settlement using Kalman filter theory. In T. Kawamoto & Y. Ichikawa (eds), *Proc. 5th Int. Conf. on Numerical Methods in geomechanics, 1–5 April, Nagoya, Japan.* Rotterdam/Boston: A.A. Balkema, 1637–1643.

Nadai, A. (1950) *Theory of flow and fracture of solids, Vol. 1*, McGraw-Hill, 207–228.

Nelder, J.A. & Mead, R. (1965) A simplex method for function minimization. *The Computer Journal*, 7, 308–313.

Nguyen, V.U. & Ashworth, E. (1985) Rock mass classification by fuzzy sets. *Proc. 26th US Symp. on Rock Mech., Vol. 2, 26–28 June, Rapid city, South Dakota, USA*, Rotterdam: A.A. Balkema, 937–945.

Noami, H., Nagano, S. & Sakurai, S. (1987) The monitoring of a tunnel excavated in shallow depth. In S. Sakurai (ed.), *Proc. 2nd Int. Symposium on Field Measurements in Geomechanics, Vol. 2, 6–9 April, Kobe, Japan*, Rotterdam: A.A. Balkema, 851–859.

Otsuka, I., Sakurai, S., Taki, H., Aoki, T., Shimo, M., Kaneko, T. & Iwano, M. (2011) Observational construction management by field measurement of large scale underground railway station by urban NATM – Railway Bosphorus tube crossing, tunnels and stations. In Q. Qian & Y. Zhou (eds), *Harmonising Rock Engineering and the Environment. Proc. 12th ISRM Congress, 18–21 October, Beijing, China*, Leiden: CRC Press/Balkema, 1769–1772.

Powell, M.J.D. (1964) An efficient method of finding the minimum of a function of several variables without calculating derivatives. *The Computer Journal*, 7, 155–162.

Rosenbrock, H.H. (1960) An automatic method for finding the greatest or least value of a function. *The Computer Journal*, 3, 175–184.

Sakurai, S. (1966) *Time-dependent behavior of circular cylindrical cavity in continuous medium of brittle aggregate*. Unpublished Ph. D. thesis, Michigan State University, East Lansing, MI, USA.

Sakurai, S. (1974) Determination of initial stresses and mechanical properties of viscoelastic underground medium. *Proc. ISRM 3rd Congress, 1–7 September, Denver, USA*. II B, 1169–1174.

Sakurai, S. (1981) Direct Strain evaluation technique in construction of underground openings. *Proc. 22nd U.S. Rock Mech. Symp., June 28–July 2, Cambridge, Mass.*: MIT, 278–282.

Sakurai, S. (1982) Monitoring of caverns during construction period. In W. Wittke (ed.), *Proc. ISRM Symposium, Rock Mechanics: Caverns and Pressure Shafts, 26–28 May, Aachen, Germany, Vol. 1*, Rotterdam: A.A. Balkema, 433–441.

Sakurai, S. (1983). Displacement measurements associated with the design of underground openings. In Kovari, A.A. (Ed.). *Proc. Int. Sympo. Field Measurements in Geomechanics, 5–8 September, Zurich, Switzerland, Vol. 2*, Rotterdam: A.A. Balkema, 1163–1178.

Sakurai, S. (1987) Interpretation of the results of displacement measurements in cut slopes. In S. Sakurai (ed.), *Proc. 2nd Int. Symposium on Field measurements in Geomechanics, 6–9 April, Kobe, Japan, Vol. 2.* Rotterdam: A.A. Balkema, 1155–1166.

Sakurai, S. (1990) Monitoring the stability of cut slopes. In R.K. Singhal & M. Vavra (eds), *Proc. 2nd Int. Sympo. Mine Planning and Equipment Selection, 7–9 November, Calgary, Canada.* Rotterdam/Brookfield: A.A. Balkema, 269–274.

Sakurai, S. (1991) Field measurement and back analysis. *Proc. 7th International Conference on Computer Methods and Advances in Geomechanics (IACMAG), 6–10 May, Cairns*, 1693–1701.

Sakurai, S. (1993) Assessment of cut slope stability by means of back analysis of measured displacements, In Pasamehmetoglu, A.G., Kawamoto, T., Whittaker, B.N. & Aydan, O. (Eds), *Assessment and Prevention of Failure Phenomena in Rock engineering, Proc. Int. Symp. on Assessment and Prevention of Failure Phenomena in Rock Engineering, 5–7 April, Istanbul, Turkey.* Rotterdam/Brookfield: A.A. Balkema, 3–9.

Sakurai, S. (1996) Practical application of back analysis in assessing the stability of underground openings. *Proc. 3rd Asian-Pacific Conference on Computational Mechanics, Vol. 4, 16–18 September, Seoul, Korea*, 2549–2564.

Sakurai, S. (1997a) Lessons learned from field measurements in tunneling. *Tunnelling and Underground Space Technology*, 12(4), 453–460.

Sakurai, S. (1997b) The monitoring of landslides by GPS. In P.G. Marinos, G.C. Koukis, G.C. Tsiambaos & G.C. Stournaras (Eds), *Engineering Geology and the Environment. Proc. Int. Symposium on Engineering Geology and the Environment, Vol. 4, 23–27 June, Athens, Greece*, 3389–3396.

Sakurai, S. (1997c) Theme lecture: Monitoring and performances in tunneling. *Proc. 14th International Conference on Soil Mechanics and Foundation Engineering, 6–12 September, Hamburg, Germany*, 2409–2412.

Sakurai, S. (1999) Interpretation of the results of displacement measurements in geotechnical engineering projects. *Proc. 5th Int. Symposium on Field Measurements in Geomechanics, 1–3 December, Singapore*, 13–18.

Sakurai, S. (2001) Back analysis of strain localization occurring in the vicinity of geostructures. In C. Desai, T. Kundu, S. Harpalani, D. Contractor & J. Kemeny (Eds), *Computer Methods and Advances in Geomechanics, Proc. 10th International Conference on Computer Methods and Advances in Geomechanics, Vol. 1, 7–12 January, Tucson, Arizona, USA*, 67–71.

Sakurai, S. (2010) Modeling strategy for jointed rock masses reinforced by rock bolts in tunneling practice. *Acta Geomechanica*, 5, 121–126.

Sakurai, S. and Akayuli, C.F.A. (1998) Deformational analysis of geomaterials considering strain-induced damage. In Cividini. A. (ed), *Application of Numerical Methods to Geotechnical Problems, Proc. 4th European Conference on Numerical Methods in Geotechnical Engineering – NUMGE98, 14–16 October, Udine, Italy*. Wien/New York: Springer, 729–738.

Sakurai, S. & Hamada, K. (1996) Monitoring of slope stability by means of GPS. *Proc. 8th Int. Symp. on Deformation Measurements, 25–28 June, Hong Kong*, 55–60.

Sakurai, S. & Hamada, K. (1997) Assessment of the stability of slopes. *Proc. 9th Int. Conference on Computer Methods and Advances in Geomechanics, 2–7 November, Wuhan, China*, 57–60.

Sakurai, S. & Ine, T. (1986) Strain analysis of jointed rock masses for monitoring the stability of underground openings. *Proc. International Symposium on Computer and the Physical modeling in Geotechnical Engineering, 3–6 December, Bangkok, Thailand*, 221–228.

Sakurai, S. & Nakayama, T. (1999) A back analysis in assessing the stability of slopes by means of surface measurements. *Proc. International Symposium on Slope Stability Engineering – IS-Shikoku'99, Vol 1, 8 11 November, Matsuyama/Shikoku, Japan*, 339–343.

Sakurai, S. & Shimizu, N. (1986) Initial stress back analyzed from displacements due to underground excavations. *Proc. Int. Symposium on Rock stress and Rock Stress measurements, 1–3 September, Stockholm, Sweden*, 679–686.

Sakurai, S & Shimizu, N. (1987) Assessment of rock slope stability by using fuzzy set theory. *Proc. ISRM 6th International Congress on Rock Mechanics, Vol. 2, 30 August–3 September, Montreal, Canada*, 503–506.

Sakurai, S. & Shimizu, N. (2006) Monitoring the stability of slopes by GPS. *Proc. Int. Sympo. Stability of Rock Slopes in Open Pit Mining and Civil Engineering Situations, 3–6 April, Cape Town, South Africa*, 353–359.

Sakurai, S. & Shinji, M. (1984) A monitoring system for the excavation of underground openings based on microcomputers. *Proc. ISRM Symposium, Design and Performance of Underground Excavations, 3–6 September, Cambridge, U.K.*, 471–476.

Sakurai, S. & Shinji, M. (2005) Back analysis of non-linear behaviour of soils and rocks considering strain-induced damage of materials. *Proc. 11th Int. Conf. of IACMAG, Vol. 1, 19–24 June, Torino, Italy*, 481–488.

Sakurai, S. & Shinji, M. (2008) Numerical simulations of laboratory experiments for determining the post-yielding mechanical parameters of soil and rock. *Proc. 12th International Conference of International Association for Computer Methods and Advances in Geomechanics (IACMAG), 1–6 October, Goa, India*, 455–461.

Sakurai, S. & Takeuchi K. (1983) Back analysis of measured displacement of tunnel. *Rock Mech. and Rock Eng.*, 16, 173–180.

Sakurai, S. & Tanigawa, M. (1989) Back analysis of deformation measurements in a large underground cavern considering the influence of discontinuity of rocks (in Japanese). *Proc. Japan Society of Civil Engineers*, No. 403/VI-10, 75–84.

Sakurai, S., Akutagawa, S. & Kawashima, I. (1993b) Back analysis of non-elastic behavior of soils and heavily jointed rocks. *Proc. 2nd Asian-Pacific Conference on Computational Mechanics, 3–6 August, Sydney, Australia*, 465–469.

Sakurai, S., Deeswasmongkol, N. & Shinji, M. (1986) Back analysis for determining material characteristics in cut slopes. *Proc. Int. Symposium on Engineering in Complex Rock Formations, 3–7 November, Beijing, China*, 770–776.

Sakurai, S., Farazmand, A. & Adachi, K. (2009) Assessment of the stability of slopes from surface displacements measured by GPS in an open pit mine. In G. Deak & Z. Agioutantis (Eds). *Sustainable Exploitation of Natural Resources Proc. 3rd International Seminar ECOMINING-Europe 21st Century, 4–5 September, Milos Island, Greece*, 239–248.

Sakurai, S., Kawashima, I. & Otani, T. (1993a) A criterion for assessing the stability of tunnels. *Proc. ISRM International Symposium, EUROCK'93, 21–24 June, Lisboa, Portugal*, 969–973.

Sakurai, S., Kawashima, I. & Otani, T. (1994b) Environmental effects on critical strain of rocks, In Balasubramaniam, A.S., Hong, S.W., Bergado, D.T., Phien-wej, N. & Nutalaya, P. (Eds), *Developments in Geotechnical Engineering, Proc. Symp. on Developments in Geotechnical Engineering, 12–16 January, Bangkok, Thailand*. Rotterdam/Brookfield: A.A. Balkema, 359–363.

Sakurai, S., Shimizu, N. & Iriyama, T. (1992) Back analysis of measured displacements in cut slopes. *Proc. ISRM Symposium: EUROCK '92, 14–17 September, Chester, UK*, 378–383.

Sakurai, S., Shimizu, N. & Matsumuro, K. (1985) Evaluation of plasti zone around underground openings by means of displacement measurements. *Proc. 5th International Conference on Numerical Methods in Geomechanics, Vol. 1, 1–5 April, Nagoya, Japan*, 111–118.

Sakurai, S., Kawashima, I., Otani, T. & Matsumura, S. (1994c) Critical shear strain for assessing the stability of tunnels. *Journal of Geotechnical Engineering, Japan Society of Civil Engineers*, 493/III–27, 185–188. (in Japanese)

Sakurai, S., Kawashima, I., Saragai, A. & Akutagawa, S. (1994a) Back analysis of the non-elastic deformational behavior of ground materials. *Journal of Geotechnical Engineering, Japan Society of Civil Engineers*, 505/III – 29, 133–140. (in Japanese)

Serata, S., Sakurai, S. & Adachi, T. (1968) Theory of aggregate rock behavior based on absolute three-dimensional testing (ATT) of rock salt. *Proc. 10th Symposium on Rock Mechanics, 20–22 May, University of Texas at Austin, Austin, Texas*, 431–473.

Shimizu, N. & Nakashima, S. (2017) Review of GPS displacement monitoring in rock engineering. In Xia-Ting Feng (ed.), *Rock Mechanics and Engineering, Vol. 3: Analysis, Modeling and Design*. Leiden: CRC Press/Balkema, 203–236.

Shimizu, N. & Sakurai, S. (1983) Application of boundary element method for back analysis associated with tunnelling problems. *Proc. 5th International Conference on Boundary Elements Methods, November, Hiroshima, Japan*, Berlin: Springer Verlag, 645–654.

Shimizu, N., Masunari, T. & Iwasaki, T. (2011) GPS displacement monitoring system for the precise measuring of rock movements. In Q. Qian & Y. Zhou (eds), *Harmonising Rock Engineering and the Environment. Proceedings of 12th International Congress on Rock Mechanics, 18–19 October, Beijing, China*. Leiden: CRC Press/Balkema, 1117–1120.

Shimizu, N., Nakashima, S. & Masunari, T. (2014) ISRM suggested method for monitoring rock displacements using the Global Positioning System. *Rock Mech. Rock Eng.*, 47, 313–328. Available from: doi:10.1007/s00603-013-0521-5. In R. Ulusay (Ed.). The ISRM

Suggested Methods for Rock Characterization, Testing and Monitoring: 2007–2014, Cham, Heidelberg, New York, Dordrecht, London: Springer, 205–220.

Stacey, T.R. (1981). A simple extension strain criterion for fracture of brittle rock. *Int. J. Rock Mech. Min. Sci. and Geomech. Abstr.*, 18, 469–474.

Terzaghi, K. & Peck, R. B. (1948) *Soil Mechanics in Engineering Practice.* New York: John Wiley & Sons.

Yang, L. & Sterling, R.L. (1989) Back analysis of rock tunnel using boundary element method. *J. Geotech. Eng. Div. Am. Soc. Civ. Eng.*, 115, 1163–1169.

Zadeh, L.A. (1965) Fuzzy sets. *Information and Control*, 8, 338–353.

Zhang D.-C., Gao X.-W. & Zheng Y. (1988) Back analysis method of elastoplastic BEM in strain space. In G. Swoboda (ed.). *Proc. 6th Int. Conf. Numer. Methods in Geomech., Innsbruck.* Rotterdam: Balkema. 981–986.

Zienkiewicz, O.C., Valliappan, S. & King, I.P. (1968) Stress analysis of rock as a 'no tension' material. *Geotechnique*, 18, 56–66.

Zienkiewicz, O.C. (1971) *The finite element method in engineering science.* London, McGraw-Hill.

References

Subject index

ISRM Book Series

Book Series Editor: Xia-Ting Feng

ISSN: 2326-6872
eISSN: 2326-778X

Publisher: CRC Press/Balkema, Taylor & Francis Group